CAMBRIDGE LIBRARY COLLECTION

Books of enduring scholarly value

Life Sciences

Until the nineteenth century, the various subjects now known as the life sciences were regarded either as arcane studies which had little impact on ordinary daily life, or as a genteel hobby for the leisured classes. The increasing academic rigour and systematisation brought to the study of botany, zoology and other disciplines, and their adoption in university curricula, are reflected in the books reissued in this series.

The Natural History of Birds

Georges-Louis Leclerc, Comte de Buffon (1707–88), was a French mathematician who was considered one of the leading naturalists of the Enlightenment. An acquaintance of Voltaire and other intellectuals, he work as Keeper at the Jardin du Roi from 1739, and this inspired him to research and publish a vast encyclopaedia and survey of natural history, the ground-breaking *Histoire Naturelle*, which he published in forty-four volumes between 1749 and 1804. These volumes, first published between 1770 and 1783 and translated into English in 1793, contain Buffon's survey and descriptions of birds from the *Histoire Naturelle*. Based on recorded observations of birds both in France and in other countries, these volumes provide detailed descriptions of various bird species, their habitats and behaviours and were the first publications to present a comprehensive account of eighteenth-century ornithology. Volume 4 covers foreign and domestic finches and flycatchers.

Cambridge University Press has long been a pioneer in the reissuing of out-of-print titles from its own backlist, producing digital reprints of books that are still sought after by scholars and students but could not be reprinted economically using traditional technology. The Cambridge Library Collection extends this activity to a wider range of books which are still of importance to researchers and professionals, either for the source material they contain, or as landmarks in the history of their academic discipline.

Drawing from the world-renowned collections in the Cambridge University Library, and guided by the advice of experts in each subject area, Cambridge University Press is using state-of-the-art scanning machines in its own Printing House to capture the content of each book selected for inclusion. The files are processed to give a consistently clear, crisp image, and the books finished to the high quality standard for which the Press is recognised around the world. The latest print-on-demand technology ensures that the books will remain available indefinitely, and that orders for single or multiple copies can quickly be supplied.

The Cambridge Library Collection will bring back to life books of enduring scholarly value (including out-of-copyright works originally issued by other publishers) across a wide range of disciplines in the humanities and social sciences and in science and technology.

The Natural History of Birds

From the French of the Count de Buffon

VOLUME 4

COMTE DE BUFFON
WILLIAM SMELLIE

CAMBRIDGE UNIVERSITY PRESS

Cambridge, New York, Melbourne, Madrid, Cape Town, Singapore,
São Paolo, Delhi, Dubai, Tokyo, Mexico City

Published in the United States of America by Cambridge University Press, New York

www.cambridge.org
Information on this title: www.cambridge.org/9781108023016

This edition first published 1793
This digitally printed version 2010

ISBN 978-1-108-02301-6 Paperback

THE NATURAL HISTORY OF BIRDS.

FROM THE FRENCH OF THE

COUNT DE BUFFON.

ILLUSTRATED WITH ENGRAVINGS;

AND A

PREFACE, NOTES, AND ADDITIONS,

BY THE TRANSLATOR.

IN NINE VOLUMES.

VOL. IV.

LONDON:

PRINTED FOR A. STRAHAN, AND T. CADELL IN THE STRAND;
AND J. MURRAY, N° 32, FLEET-STREET.

MDCCXCIII.

CONTENTS

OF THE

FOURTH VOLUME.

 The

CONTENTS.

The

CONTENTS.

CONTENTS.

CONTENTS.

The

CONTENTS.

11. The

CONTENTS.

5. The

CONTENTS.

The

CONTENTS.

14. The

CONTENTS.

THE

THE

NATURAL HISTORY

OF

BIRDS.

The CANARY FINCH*.

Le Serin des Canaries, Buff.
Fringilla Canaria, Linn.
Serinus Canarius, Briff. Ray, Will. and Kolb.
Paſſere di Canaria, Zinn.
Canarien Vogel, Wirs.

IF the Nightingale be the ſongſter of the grove, the Canary Finch is the muſician of the chamber. The melody of the former is derived from Nature alone, that of the latter is directed and improved by our inſtructions. With a weaker voice, with leſs extent of modulation, with leſs variety of notes, the Canary Finch has a finer ear, greater facility of imitation †, and a more

* The Linnean character :—"The bill and body yellowiſh-white, the feathers of the wings and tail greeniſh, the bill whitiſh." It is four inches and a half long.

† The Canary learns to ſpeak, and articulate many little names very diſtinctly. . . . By means of a flageolet, it can acquire two or

 three

a more retentive memory; and, as the characters of animals depend principally on the quality of their perceptions, this delicate bird, alive to every impreſſion, becomes alſo more ſocial, more gentle; forms acquaintance, and even ſhews attachment*. Its careſſes are amiable, its little pets are innocent, and its anger neither hurts nor offends. Its habits too approach nearer to our own; it feeds on grain, like the other domeſtic birds; it is more eaſily bred than the Nightingale, which lives only on fleſh and inſects, and which requires its meals to be purpoſely prepared. Its education is attended alſo with greater ſucceſs; it readily lays aſide the melody of its native airs to adopt the harmony of our voices and inſtruments; it eagerly follows the notes, and improves and heightens their delicacy. The Nightingale, proud of its independent warble, ſeems deſirous to preſerve its purity; at leaſt, he ſlights our muſic, and can hardly be brought to learn a few ſongs. The Canary Finch prattles or whiſtles; but the Nightingale deſpiſes what he deems the perverſion of his talents, and perpetually recurs to the rich beauties of Nature.

three airs, which it chants in their proper tone, always keeping due time, &c. *Traité des Serins des Canaries, par Hervieux*, 1713. A Canary, which, when young, was placed very near my deſk, got an odd ſort of ſong; it imitated the ſound made by telling crowns. *Note communicated by Hebert, receiver general at Dijon.*

* It becomes ſo familiar and ſo fond of careſſing, that a thouſand and a thouſand times it comes to kiſs and bill its maſter, and never fails to anſwer his call. *Traité des Serins, par M. Hervieux.*

His

His ever-varied ſong can never be altered by man; that of the Canary Finch is more pliant, and can be modelled by our taſte. The one therefore contributes more than the other to the comforts of ſociety; the Canary Finch ſings at all times, recreates our ſpirits in the gloomieſt weather, and even adds to our happineſs; it amuſes all young people, and is the delight of the recluſe; it relieves the languors of the cloiſter, and infuſes cheerfulneſs into innocent and captive minds; and its little loves, which are manifeſt when it breeds, have a thouſand and a thouſand times awakened the tenderneſs of feeling hearts. It is as uſeful, as the Vulture is pernicious.

To the happy climate of the Heſperides, this charming bird owes it birth, or, at leaſt, its perfection: for, in Italy *, there is a ſpecies ſmaller than that of the Canaries, and in Provence, another almoſt as large †; but both theſe are more

* *Citrinilla*, Geſner; *Vercellino*, Olina.—" Sparrow variegated " above with yellowiſh green; below dilute greeniſh; the wing " and tail quills blackiſh, the outer margin greeniſh." The Italian *Serin*. Brisson. It is the *Fringilla Citrinella* of Linnæus, and the *Citril Finch* of Latham. Its ſpecific character:—" It is ſomewhat greeniſh, its back ſpotted with duſky, its legs fleſh co- " loured."

† *Serinus*, *Serin*, *Cenicle*, *Cereſin*, *Cinit*, *Cedrin*.—" Sparrow " varying above with greeniſh yellow, below dilute greeniſh, the " ſides marked with duſky longitudinal ſpots, a greeniſh yellow " ſtripe on the wings; the quills of the wings and of the tail " duſky above, the outer margins grayiſh-green, the tips whitiſh." The *Serin*. Brisson. It is the *Fringilla Serinus* of Linnæus, and the *Serin Finch* of Latham. Its ſpecific character:—" It is ſomewhat " greeniſh,

more ruſtic, and may be regarded as the wild branches of a poliſhed ſtem. All the three intermix in the ſtate of captivity; but in the range of nature, each propagates in its peculiar climate. They are therefore permanent varieties, and ought to be diſtinguiſhed by ſeparate names. The largeſt was called *Cinit*, or *Cini*, in the time of Belon (above two centuries ago); and in Provence it is termed at preſent *Cini*, or *Cigni*, and the Italian kind *Venturon*. I ſhall diſtinguiſh theſe three varieties by the epithets *Canary*, *Cini*, and *Venturon*.

The *Venturon*, or the Italian Bird, is found not only through the whole of Italy, but in Greece *, Turkey, Auſtria, Provence, Languedoc, Catalonia, and probably in all climates of that temperature. Some years, however, it is very rare in the ſouthern provinces of France, and particularly at Marſeilles. Its ſong is pleaſant and varied. The female is inferior to the male both in the charms of its notes and in the beauty of its plumage. The ſhape, the colour, the voice, and the food of the Venturon and of the Canary, are nearly the ſame; and the only difference is,

" greeniſh, its lower mandible whitiſh, its back and ſides ſpotted " with duſky; a yellow ſpot on the wings."

* The ancient Greeks called this bird Τραυπις; and the modern Greeks, according to Belon, Σπινιδυα. The Turkiſh name is *Sare*: in ſome parts of Italy it is termed *Luguarinera*, *Beagana*, *Raverin*; in the neighbourhood of Rome, *Verzellino*; in Bologna, *Vidarino*; at Naples, *Lequilla*; at Genoa, *Scarino*; in the Trentin, *Citrinella*; in Germany, *Citrynle* or *Zitrynle*; at Vienna, *Citril*.

that

that the Italian bird is ſenſibly ſmaller, and its warble neither ſo clear nor ſo melodious.

The *Cini*, or Green Bird of Provence, is larger than the Venturon, and its tones are much fuller. It is diſtinguiſhed by the beauty of its colours, the loudneſs of its ſong, and the variety of its notes. The female, which is rather larger than the male, and has fewer yellow feathers, chants not like its mate, and anſwers only in monoſyllables. It feeds on the ſmalleſt ſeeds it can find in the field; lives long in a cage, ſeems fond of lodging with a gold-finch, whoſe accents it ſeems attentively to hear, and adopt, to vary its own warble. It occurs not only in Provence, but alſo in Dauphiné, in the Lyonnois *, in Bugey, in Geneva, in Switzerland, in Germany, in Italy, and in Spain †. It

* I have ſeen in the country in Bugey, and near Lyons, birds very like the Canary Finches, and they were called *Signis* or *Cignis*. I have alſo ſeen them at Geneva in cages, and their ſong did not ſeem to me very pleaſant.—I think they are called at Paris "the Swiſs Serins." *Note of* M. HEBERT.

"The German Serins are much commended; they excel the Canary Finches by their beauty and their ſong. They are not inclined to grow fat; the force and extent of their warble prevent, it is ſaid, that effect. They are raiſed in cages or in chambers fitted up for them, and having an eaſterly aſpect; they hatch thrice a-year, from the month of April to Auguſt." *Le Parfait Oiſeleur*.

This is not altogether accurate; for the ſong of theſe German Serins, which are the ſame with thoſe of Switzerland and Provence, though full and ſhrill, is far from having the ſweetneſs and mellowneſs of the Canary Birds.

† In Catalonia, it is called *Canari de Montanya*; in Italy, *Serin*, or *Scarzerin*; in Germany, *Fædeule*; in the neighbourhood of Vienna, *Hirn-gryll*; in Switzerland, *Schwederle*.

is the ſame bird that is called in Burgundy the *Serin*; it builds among the oſiers planted along the ſides of the rivers, and its neſt is lined with hair, and caſed with moſs. This bird, which is pretty common in the vicinity of Marſeilles, and in the ſouth of France as high as Burgundy, is unfrequent in the northern provinces. Lottinger ſays it is only migratory in Lorraine.

The prevailing colour of the Venturon as well as that of the Cini, is a green yellow on the upper part of the body, and a yellow green on the belly; but the Cini, larger than the Venturon, is diſtinguiſhed by the brown colour which appears in longitudinal ſpots on the ſides of the body, and in waves below *; whereas in our

* I ſhall here inſert an excellent deſcription of the Cini, which M. Hebert ſent to me. "This bird is ſomewhat ſmaller than the Canary Finch, which it much reſembles. It has preciſely the ſame plumage with a kind of Canary called the *Gray Canary*, which is perhaps the original bird, and the varieties are owing to domeſtication.

"The fore-part of the head, the orbits of the eyes, the under-part of the head, a ſort of collar, the breaſt and belly as far as the toes, are of a jonquil-colour, with a tinge of green. The ſides of the head, and the higher part of the wings, are mixed with green, jonquil, and black. The back, and the reſt of the wings, are daſhed with green, gray, and black. The rump is jonquil. The breaſt, though of a ſingle colour, is waved. The ſpots ſcattered on the plumage of the Cini are not diſtinctly marked, but run into each other; thoſe on the head are much finer, and like points; on the two ſides of the breaſt, and under the belly parallel to the wings, are ſpots or ſtreaks of black.

"The tail is forked, conſiſting of twelve quills; the wings are of the ſame colour with the back; the ends of the coverts at the origin of the great quills, are ſlightly edged with a ſort of dull yellow;

our climate, the common colour of the Canary is uniform, and of a citron-yellow on all the body, and even on the belly: it is only the tips of the feathers however that are tinged with that delicate hue, the reſt of them being entirely white. The female is of a paler yellow than the male; but this citron-colour verging more or leſs on white, which the Canary aſſumes in our climate, is not the tint of its native climate, for it varies according to the different temperatures. " I have obſerved," ſays one of our moſt intelligent naturaliſts, " that the Canary Finch, " which becomes entirely white in France, is in " Teneriffe of almoſt as deep a gray as the lin- " net; this change of colour is owing probably " to the coldneſs of our climate." The colour can be varied alſo by the difference of food, by confinement, and above all by the mixture of

low; the great quills and the tail are ſimilar, and of a brown verging to black, with a ſlight edging of gray; the tail is ſhorter than that of the Canary bird.

" In general this bird is jonquil below, and variegated on the back with different colours, in which the green predominates, though we cannot aſſert that this is the ground colour; on the back not a ſingle feather but is variegated with ſeveral colours.

" The bill is pretty much like that of the Canary, but rather ſhorter and ſmaller; the upper mandible is level with the crown of the head, has little concavity, broader at its baſe, and ſcalloped near its origin; the lower mandible is more concave, placed diagonally under the upper, into which it is encaſed.

" This Cini was only two inches and ſeven lines from the top of the head to the origin of the tail, which was only one inch and ten lines; the wings extend to the third of the tail; the legs are very ſlender; the tarſus ſix lines long, and the toes near as much; the nails are not regularly hooked."

 breeds.

breeds. In the beginning of this century the curious in birds reckoned already twenty-nine varieties of Canaries, and all of them were diſtinctly deſcribed * The primitive ſtock from which theſe were deſcended is the Common Gray Canary. All thoſe which have other uniform colours derive them from the difference of climates. Thoſe which have red eyes are more or leſs inclined to a pure white plumage; and the variegated are rather artificial than natural †.

But,

* *I ſhall here enumerate them all, beginning with thoſe which are moſt common:—*

1. The Common Gray Canary.
2. The Gray Canary, with down and white feet. *Variegated breed.*
3. The White-tailed Canary. *Variegated breed.*
4. The Common Flaxen Canary.
5. The Flaxen Canary, with red eyes.
6. The Golden Flaxen Canary.
7. The Flaxen Canary, with down. *Variegated breed.*
8. The White-tailed Flaxen Canary. *Variegated breed.*
9. The Common Yellow Canary.
10. The Yellow Canary, with down. *Variegated breed.*
11. The White-tailed Yellow Canary. *Variegated breed.*
12. The Common Agate Canary.
13. The Agate Canary, with red eyes.
14. The White-tailed Agate Canary. *Variegated breed.*
15. The Agate Canary, with down. *Variegated breed.*
16. The Common Pink Canary.
17. The Pink Canary, with red eyes.
18. The Golden Pink Canary.
19. The Pink Canary, with down. *Variegated breed.*
20. The White Canary, with red eyes.
21. The Common Variegated Canary.
22. The Variegated Canary, with red eyes.
23. The Flaxen Variegated Canary.
24. The Flaxen Variegated Canary, with red eyes.
25. The Black Variegated Canary.
26. The Jonquil-black Variegated Canary, with red eyes.
27. The Jonquil-black Variegated Canary, of a regular plumage.
28. The Full Canary, (that is entirely of a jonquil-yellow,) which is the rareſt.
29. The Creſt Canary, (or rather the Crowned,) which is one of the moſt beautiful.

Traité des Serins de Canaries, par Hervieux, 1713.

† "The ſhades and arrangement of the colours of the Variegated Canaries differ exceedingly; ſome are black on the head, others not; ſome are ſpotted irregularly, and others with great regularity. The differences of colour are commonly perceived only on the upper part of the bird; they conſiſt of two large black ſpots on

But, besides these primary varieties, which seem to have been introduced by the change of climate, and some secondary ones that have since appeared, there are others more apparent, and which result from the union of the Canary with the Venturon, and with the Cini; for not only do these three birds breed together, but the young hybrids are capable of procreation. The same may be said in regard to the fruits of the commerce of the Canary with the Siskin, with the Goldfinch, with the Linnet, with the Yellow-hammer, with the Chaffinch, and even it is said with the Sparrow*. These birds, though very different from each other, and apparently discriminated widely from the Canaries, can by proper care and attention be made to pair with them. The Canary must be removed from those of its own species; and the experiment seems to succeed better when performed with the female than with the male. The Siskin and Goldfinch are the only birds with which, it is

on each wing, the one before and the other behind, in a large crescent of the same colour placed on the back, pointing its concavity towards the head, and joining by its horns to the two anterior black spots of the wings. Lastly, the tail is surrounded behind by an half-collar of gray, which seems to be a compound colour resulting from the intimate mixture of black and yellow. The tail and its coverts are almost white." *Description des Couleurs d'un Canari Panaché, observé avec M. de* MONTBEILLARD.

* D'Arnault assured Salerne that he saw at Orleans a gray hen Canary which had escaped from the volery, couple with a sparrow, and make her hatch in a *sparrow-can*, which thrived. *Amusemens Innocens, ou le Parfait Oiseleur*, 1774.

well

well aſcertained, that the male Canary can propagate. On this ſubject one of my friends thus writes me; a perſon of as much experience as veracity*.

"For thirty years paſt I have raiſed many of theſe ſmall birds, and have paid particular attention to breeding them; I can therefore from long obſervation aſſert the following facts. When we wiſh to pair Canaries with Goldfinches, we muſt take the Goldfinches at ten or twelve days old, and put them in the neſt with Canaries of the ſame age; we muſt raiſe them together in the ſame volery, and accuſtom them to the ſame food. It is uſual to lodge the male Goldfinches with the female Canaries, and they aſſociate much more readily than if the female Goldfinches were joined with the male Canaries. We may obſerve however, that the union is more tardy, becauſe the Goldfinch is later in pairing than the Canary; on the contrary, if the female Goldfinch be placed beſide the male Canary they ſooner couple †. To ſucceed we muſt never admit the male Canary into the voleries where there are female Canaries, for he would then prefer them to Goldfinches ‡."

* Father Bougot.

† This proves (as we ſhall afterwards obſerve) that the female is not ſo much ſtimulated to love by nature, as rouſed by the ſolicitation of the male.

‡ Geſner ſays, that a Swiſs bird-catcher trying to pair a cock Canary with a hen *Sarzerine,* (Cini,) had eggs, but that theſe were addle.

"With

"With regard to the union of the male Canary with the female Siſkin, I am certain that it ſucceeds well. I have kept for nine years in my volery a female Siſkin, which never failed to make three hatches annually for the firſt five years, though the number was reduced to two in the four following years. I have other Siſkins which have bred with the Canaries, without being raiſed or kept ſeparately. We need only leave the male or female Siſkin in the chamber with a good number of Canaries, and we ſhall perceive them pair with the Canaries, at the ſame time that theſe pair with each other; whereas, to ſucceed with the Goldfinches, they muſt be ſhut up in a cage with the Canary, and every bird of the ſame ſpecies muſt be excluded. The Siſkin lives as long as the Canary, and eats the ſame food with much leſs reluctance than the Goldfinch.

"I have alſo put Linnets with Canaries; but there is ſeldom ſucceſs, unleſs we lodge the male Linnet with the female Canary; for the hen Linnet makes no neſt, but drops her eggs in the cage, and theſe are generally addle. I have made the experiment with them, having placed them under female Canaries, and frequently found that they did not hatch.

"It is very difficult to pair Chaffinches and Yellow-hammers with Canaries. I have kept for three years a female Yellow-hammer with a male Canary, and it has laid only addle eggs.

eggs. This has alſo been the caſe with the female Chaffinch; but when the female Canary is covered by the male Chaffinch and Yellow-hammer ſhe has prolific eggs."

From theſe facts, and ſome others which I have collected, it appears that the Siſkin is the only bird of which the male and female propagate equally with thoſe of the male or female Canaries. The female Canary alſo breeds readily with the Goldfinch; not ſo eaſily with the cock Linnet; and even produces, though with more difficulty, under the male Chaffinches, Yellow-hammers, and Sparrows; whereas the male Canaries cannot impregnate the females of theſe laſt. Nature is therefore more uniform and permanent in the male than in the female; in the former the characters are deeply imprinted; in the latter, the qualities are modified by the influence of external cauſes. In the few experiments which I have made on the union of ſome contiguous ſpecies of quadrupeds, I obſerved that the ewe eaſily bred under the he-goat; but that the ram could not propagate upon the ſhe-goat. I have been told of an inſtance in which a ſtag covered a cow; but the bull has been known to copulate with the hind. The mare breeds more readily with the jack-aſs, than the ſhe-aſs with the ſtallion. In general, mules partake more of the characters of the female than of the male, whoſe features are more ſtrongly marked.—Theſe facts correſpond with thoſe we have

have mentioned with regard to the croſs-breed of birds. It is evident that the female Canary is much more related than the male to the ſpecies of the Yellow-hammer, of the Linnet, of the Chaffinch, and of the Sparrow; ſince it breeds with all theſe, while the male will not. I uſe the term *will*, for perhaps it is only from a want of reſolution that the female yields to the ſolicitations of thoſe of a different ſpecies: however, an examination of the products of this intermixture affords concluſions that agree with all that I have ſaid of the generation and developement of animals; and as the ſubject is important, I ſhall here ſtate the principal facts.

The firſt variety which ſeems to conſtitute two diſtinct families in the ſpecies, conſiſts of the variegated Canaries, and thoſe whoſe plumage is uniform. The white ſort, or the yellow-citron, are never variegated; only when the latter are four or five years old, the tips of the wings and of the tail become white. The gray kind are not of an uniform colour; ſome feathers are affected by different ſhades, and ſome individuals are of a lighter or of a darker tinge. The agate is ſpread uniformly, though it varies in the intenſity. The pink coloured are more ſimilar, the tint being uniform, not only in different individuals, but in different parts of the ſame individual. In thoſe which conſiſt of ſeveral colours, the Yellow Jonquil ſort are variegated with blackiſh, and are commonly black on the

the head. In ſome Canaries, the plumage is tinctured with all the ſimple colours which we have mentioned; but the Yellow Jonquil are the moſt variegated with black.

When Canaries are paired of an uniform colour, that of their young is alſo uniform. If both parents are gray, for inſtance, the progeny is alſo commonly gray; and this is the caſe with the pink, with the white, with the yellow, and with the agate. If the parents be of different colours, the fruits of their commerce will have a richer plumage; and as the combinations that can take place are numerous, an immenſe variety of tints may be produced. But when the Canaries of an uniform colour are croſſed with thoſe which are variegated, the changes are prodigiouſly multiplied, and there is ſcarce any limit to the ſportive modifications. Nay, it often happens that parents of an uniform colour have beautiful variegated offspring, derived from the mixture of ſhades in themſelves or their progenitors *.

With reſpect to the intercourſe of the Canary with birds of other ſpecies, I ſhall here ſtate the obſervations which I have collected. Of all

* To have beautiful birds, we muſt pair the flaxen variegated male with a white-tailed yellow female; or at leaſt a variegated male with a white-tailed flaxen female, or any other female that is white-tailed, except only the gray. And when we want to obtain a fine Jonquil, we ſhould ſet a black variegated male with a white-tailed yellow female. *Amuſemens Innocens.*

the

the Canaries, the Cini or Green Canary has the ſtrongeſt voice, and appears to be the moſt vigorous, and the moſt ardent in propagation ; it is ſufficient for three female Canaries, and while they are ſitting in the neſts, it ſupplies them and their young with food. The Siſkin and the Goldfinch are neither ſo powerful nor ſo aſſiduous, and a ſingle female Canary ſatisfies their appetite.

The croſs-breed of the Cini, of the Siſkin, and of the Goldfinch, with the hen Canary, are ſtronger than the Canaries, ſing longer, and their notes are fuller and more ſonorous ; but they are ſlower in receiving inſtruction : for the moſt part, they whiſtle imperfectly, and ſeldom one can be found that can repeat a ſingle air complete.

When we wiſh to have a breed of the Goldfinch with the hen Canary, the former ſhould be two years old, and the latter one, for the Canary arrives ſooner at maturity. It will be better if both be reared together, though that precaution is not abſolutely neceſſary; and the Author of " the Treatiſe on Canaries" is miſtaken in aſſerting, that the hen muſt never have had commerce with a male of her own ſpecies, and that this would prevent her from receiving thoſe of a different kind. The following fact is directly contrary to this opinion : " I happened," ſays Father Bougot, " to put twelve Canaries together, four males and eight females.

Bad chickweed killed three of theſe males, and all the females loſt their firſt hatch. In the ſtead of theſe Cock-Canaries, I thought of ſubſtituting three male Goldfinches which I caught, and put them into the volery in the beginning of May. I had, towards the end of July, two neſts of young, which thrived as well as poſſible; and in the following year, I had three hatches of each cock Goldfinch with the female Canaries. Theſe commonly do not breed with the Goldfinch, except from the age of one to four; while they continue prolific with their own males till eight or nine: and it is only the variegated female that breeds with the Goldfinch after its fourth year. We muſt never put the Goldfinch into a volery, for it demoliſhes the neſts, and breaks the eggs of the other birds." It appears then that the hen Canaries, though accuſtomed to the commerce of their own ſpecies, liſten to the invitation of the male Goldfinches, and ſubmit without reluctance to their embraces. The union is even as productive as with their proper mates, ſince they have three hatches a-year with the Goldfinch. The caſe is different in the intercourſe of the cock Linnet with the Canary, there being commonly but one hatch, and very ſeldom two in the year.

The progeny of the Canaries with the Siſkins, with the Goldfinches, &c. are prolific, and can breed not only with both the ſpecies from which they

they ſprung, but likewiſe with each other; and thus may perpetuate an endleſs variety *. We muſt own, however, that the fruits of the intermixture in theſe hybrids are far from being ſo certain or ſo numerous as in the pure ſpecies; they generally have only one hatch a-year, ſeldom two, their eggs are often addle, and the ſucceſs depends on many minute circumſtances, which it would be impoſſible to obſerve, and ſtill leſs to deſcribe with preciſion. It is ſaid, that in theſe there are always more males than females. "A hen Canary (ſays Father Bougot) and a Goldfinch have, in the ſame year at three different times, laid me nineteen eggs, which all hatched, and of the young were only three females." It would be proper to aſcertain this fact by repeated obſervations. In the pure breeds of ſeveral birds, as in the partridges, it has alſo been remarked that the males exceed the females. The ſame remark applies to the human race: in our climates, ſeventeen boys are born for ſixteen girls. We know not the preciſe proportion between the male and female partridges, but that the former are more numerous than the latter we are certain, becauſe in the pairing

* Sprengel has made many obſervations with regard to the croſs-breeds of the Canaries with other ſpecies, and particularly with the Goldfinches; he has ſhewn that the progeny continued to propagate with each other, and with the parent races. The proof which he has adduced is complete, though before him theſe hybrids were conſidered as ſterile. *Amuſemens Innocens*.

 ſeaſon,

ſeaſon, there are always cock birds that want mates. It is likely, however, that ſixteen to three, as in the croſs-breed of the Canary and Goldfinch, is a greater inequality than ever takes place in a pure breed. I have been told, that the ſhe-mules, got between the aſs and the mare, exceed in number the he-mules; but I could never obtain accurate information on that ſubject. Our buſineſs then is to determine by obſervation the number of males and of females in the pure race of the Canary, and afterwards to examine if that of the males be ſtill greater in the croſs-breed of the Goldfinch and hen Canary. What diſpoſes me to entertain this opinion is, that the character of the male is in general more deeply impreſſed on the progeny, than that of the female. Theſe hybrids, which are ſtronger than the Canaries, and have a louder voice, are alſo longer lived. But there is an obſervation which applies alike to all, that the period of their lives is abridged by their ardour in propagation. A cock Canary raiſed by itſelf, and without intercourſe with the female, lives generally thirteen or fourteen years; and the croſs-breed of the Goldfinch will reach its eighteenth or nineteenth year. The croſs-breed of the Siſkin, if kept apart from the females, live fifteen or ſixteen years. Whereas the cock Canary, which has one or ſeveral females, ſeldom lives longer than ten or eleven years; the croſs-breed of the Siſkin eleven or twelve, and that of the Goldfinch fourteen or fifteen. It is alſo neceſſary to part them from the females after the time of hatching; that is, from the month of Auguſt to that of March; elſe the heat of their paſſion would waſte them, and abridge their lives by two or three years.

To theſe remarks, which are all intereſting, we may ſubjoin a general and more important obſervation, which may throw ſome light on the generation of animals, and the developement of their different parts. It has been conſtantly noticed in the copulation of the Canaries, both with thoſe of their own ſpecies and with thoſe of other races, that the offspring reſembled the father in the head, the tail, and the legs, and the mother in the reſt of the body. The ſame has been obſerved in quadrupeds. The mule got between the jack-aſs and the mare, has the thick body of the former, and the ears, the tail, and the thin legs of the latter. It appears therefore that, in the mixture of the two ſeminal liquors, however intimate we ſuppoſe it to be, the organic molecules furniſhed by the female occupy the centre of that living ſphere which increaſes in all dimenſions, and that the molecules injected by the male ſurround and incloſe theſe; ſo that the extremities of the body proceed more imdiately from the father than from the mother. The ſkin, the hair, the colours, which may be conſidered as the exterior of the body, retain moſt of the paternal impreſſions. In the croſs-

 breed

breed which I obtained from the copulation of the he-goat with the ewe, they all had, inſtead of a ſoft fleece, the ſhaggy coat of the male. In the human ſpecies, we may generally perceive, that the ſon reſembles the father more than the mother, in his legs, his feet, and his hands; in his writing, in the quantity and colour of his hair, in his complexion, and in the bulk of his head: and the mulattoes born of a negreſs loſe more of the dark tinge than thoſe of a white woman. All theſe facts ſeem to confirm our general idea, that the female molecules occupy the centre of the *fœtus*, and, though brought into the cloſeſt union, are more abundant near the ſurface.

In general, the beauty of the ſpecies can never be improved or even preſerved, without croſſing the breed: and the elegance of form, the ſtrength and vigour of the body, depend almoſt ſolely on the proportion of the limbs. Accordingly, it is the males alone, which in man and in the animals ennoble the race. Large, generous mares, covered by ſorry little horſes, have always ill-made colts; but a fine ſtallion will get a beautiful progeny from even the uglieſt mares, and the more diſſimilar are the kinds of the parents, the handſomer will be the offspring. The ſame may be ſaid of ſheep: the breed is always improved by ſtrange rams, while the ordinary ſmall ſort can never get good lambs from the moſt excellent ewes. I could enlarge upon this important ſubject,

ſubject, but I ſhould make too long a digreſſion. Yet, to detail all the facts reſpecting the intermixture of animals, would be the moſt real ſervice that can be rendered to Natural Hiſtory. As many people employ or amuſe themſelves in breeding Canaries, which requires little time, numerous experiments might be made by croſſing them, and continuing to mix the fruits of the embrace. I am confident that, by combining theſe obſervations with thoſe upon the quadrupeds and upon man, we might be able to eſtimate the preciſe influence of the male in generation, compared with that of the female; and conſequently, from the general relations, to pronounce whether ſuch a male would ſuit ſuch a female, &c.

But in the quadrupeds, as well as in man and in the ſmall birds, the difference of the moral qualities often diſturbs the correſpondence of the phyſical properties. If any thing could prove that the character of the individual is an original impreſſion of nature which education can never alter, it is an inſtance in the Canaries. "They "almoſt always," ſays Hervieux, "differ from "each other in their tempers; ſome males are "always ſad, and, as it were, abſorbed in re-"veries, generally bloated, and ſing but ſel-"dom, in a mournful tone.... require an "immenſe time to learn, are imperfectly ac-"quainted with what they are taught, and eaſily "forget it.... Theſe Canaries are often ſo ſlo-

 "venly,

" venly, that their feet and tail are always dirty; " they never gain the affection of their females, " which they soothe not with their song, while " engaged in hatching; and the young are " little better than their fathers. . . . There are " other Canaries which are so wicked, that they " kill the female; the only way to succeed is, " to give them two females, which will unite in " their common defence; and after subduing " their mate by force, they will retain the do- " minion by love*. Others are of a disposi- " tion so barbarous, that they break the eggs " and eat them; or if these have escaped their " ferocity, they lay hold of the callow brood by

* " Sometimes the pravity of their disposition is in a certain measure compensated by other qualities; such, for instance, as their melodious song, their beautiful plumage, and their familiar turn. If you would wish to make them breed, you must give them two strong females one year older than themselves; and the females should be accustomed for several months previous to live in the same cage, that they may have no jealousy to each other. And a month before the love-season, they must be both set together in the breeding cage, and at the proper time the male should be let in among them. He will be very quarrelsome the first three days, but the females, uniting against him, will certainly in the end gain the ascendant, and he will be obliged to submit, and at last become attached to them. These kind of forced marriages often succeed better than others from which more might be expected. To preserve the hatch, the first egg should be removed, and an ivory one put in its place; the same must be done the following days, always taking them away as fast as they are laid, lest the male should break them; and after the last egg is dropped, the male must be shut in a separate cage, and the female permitted to hatch undisturbed. After the young are ready to be taken from the mother, the prisoner may be returned to his female."

Traité des Serins des Canaries.

" the

" the bill, drag them into the cage and murder " them *." Some are ſo wild and independent, that they will not ſuffer themſelves to be touched or careſſed, and can neither be governed nor treated like the reſt: they ſpurn at the leaſt interference, and it is only when left to the impulſe of their humours that they will couple and breed. Others are exceſſively indolent; the gray for inſtance ſcarcely ever take the pains to build a neſt, but it muſt be provided for them, &c. All theſe characters are, we ſee, very different from each other, and from thoſe of our favourite Canaries, which are ever joyous, and ever chanting; are

* " There are males of a weak habit, indifferent about the females, and always ſick after neſtling; theſe muſt not be paired, for I obſerve that the offspring reſemble the father. There are others ſo libidinous, that they drive the female from her neſt, and would not allow her to ſit; theſe are of a hardy conſtitution, have a ſuperior ſong, a finer plumage, and are tamer. Others break the eggs, and kill the young, the more to enjoy the female. Others ſhew a predilection for an individual, and will, out of twenty, ſelect their favourite, to which they will pay particular attention. Thoſe have a good temper, and will communicate it to their progeny. Others diſcover fondneſs for no female, and remain inactive and unproductive. The ſame difference of character and temperament are to be found in the females. The jonquil females are moſt gentle; the agate ſort are capricious, and often deſert their young to join the male; the variegated females are aſſiduous on their eggs and affectionate to their young, but the variegated males are the moſt ardent of all the Canaries, and muſt have two or three females, elſe they will drive them from the neſt and break the eggs. Thoſe which are entirely jonquil have nearly the ſame fire of temper, and require two or three females. The agate males are the feebleſt, and the females of that kind often expire upon their eggs." *Note communicated by Father* Bougot.

ſo tame and ſo lovely; are excellent huſbands, and affectionate fathers; are of ſo mild a temper, and of ſo happy a diſpoſition, that they receive every generous impreſſion, and glow with each exalted feeling. They continually amuſe the female by their ſong, they ſooth the languor of her occupation, they entreat her to take relief in hatching, and, in her place, they ſit ſeveral hours every day; they alſo feed the young; and laſtly, receive whatever inſtructions are given. From theſe alone we are to judge of the ſpecies, and I mention the others only, to ſhew that, even in animals, the temper is derived from nature, and not formed by education.

Moreover, the apparently wicked diſpoſition, which drives them to break their eggs and kill their young, proceeds often from the fire of their amorous paſſions. To enjoy the female oftener, and riot in the fulneſs of pleaſure, they plunder the neſt, and deſtroy the deareſt objects of their affection. The beſt way to breed from theſe birds, is to ſeparate them, and put them in the cage; it will be much better to give them a chamber having a ſunny aſpect, and facing the eaſt in winter. For in the cage they will break the eggs to repeat their embrace; but when they are lodged in an apartment where there are more females than males, they will pay their addreſſes to another, and allow the firſt to hatch undiſturbed. Beſides, the males, from jealouſy, will

not

not ſuffer diſorders to be committed; and when one is prompted by ardor to teaſe his female and break her eggs, they give him a ſound beating, ſufficient to mortify his concupiſcence.

The materials given to build their neſt, are the ſcrapings of fine linen rags, cows and ſtags hair, uſeleſs for other purpoſes, moſs, and ſlender dry ſtalks of hay. The Goldfinches and Siſkins, when lodged with the hen Canaries, to obtain a croſs-breed, make uſe of the moſs and hay, but the Canaries prefer the hair and lint; but theſe muſt be well divided, leſt the fibres, ſticking to their feet, ſhould occaſion the eggs to be broken.

To feed them, a crib is placed in the chamber, which is pierced all round with holes that admit the head; and into it a portion of this mixture is put; three pints of rape-ſeed, two of oats, two of millet, and laſtly, a pint of hemp ſeed, and the crib is repleniſhed every twelve or thirteen days, taking care that the grains be well cleaned and winnowed. This food is proper when they are only ſitting; but the day before the young are excluded from the ſhell, the parents ought to have a dry cake baked without ſalt, and after eating it, they ſhould have hard boiled eggs; one being ſufficient for two males and four females, and two for four males and eight females, and ſo on in proportion. They ought to have no ſallad or greens while rearing the offspring, for this would weaken much the young. But to vary ſomewhat their diet, they ſhould every three

three days be presented on a plate with a bit of white bread soaked in water, and squeezed in the hand; this being not so rich as the cake, and preventing them from growing too fat while breeding. It would also be proper to give them a few Canary seeds *, but only once in two days, lest they be heated too much. Sugar biscuit commonly produces that effect, and is attended with another still worse, that the hens fed on it lay eggs that are addle, or too small and tender. When the eggs are small, the rape-seed should be boiled every day to blunt its acrimony. "Long experience," says Father Bougot, "has informed us, that this sort of food "agrees best, whatever the authors, who have "written on Canaries, may assert."

After the hatch, the Canaries ought to be purged with plantain and lettuce-seeds; but care must be taken to remove the young birds, which would be greatly weakened by this regimen, and the parents must not be confined to it longer than two days. When you want to feed them with the stick, you ought not, as most bird-catchers advise, leave them with the mother till the eleventh or twelfth day; you ought to remove them with their nest as early as the eighth day. The food for the young

* *Alpis* in French, the Canary-grass being termed *Alpiste*. It is the *Phalaris Canariensis* of Linnæus. It is a native of the Canary islands, whence it springs spontaneously in the corn fields; but is now cultivated in small quantities in many parts of Europe.

Canaries

Canaries ought to be prepared before hand; it is a paſte compoſed of boiled rape-ſeed, yolks of eggs and crumbs of cake mixed together, and beaten up with a little water. It ought to be given to them every two hours, and rammed into their bill. It muſt not be too liquid, leſt it turn ſour, and it muſt be made freſh every day till the young can eat without aſſiſtance.

The produce of theſe captive birds is not ſo regular, but appears to be more numerous than it probably would be in the ſtate of liberty. Some females have four or five hatches annually, and lay four, five, ſix, or even ſeven eggs each time; and generally they have three hatches, and the moulting hinders them from another *. Some however ſit even during moulting, if they happen to be laying before that ſeaſon. The birds of the ſame hatch do not all moult at the ſame time. The weakeſt firſt drop their feathers, and the ſtrongeſt more than a month afterwards. In Jonquil Canaries this change

* "There are females which never lay at all, and are called *brehaignes* (barren); others lay only once or twice in the whole year, and even repoſe two or three days between the firſt and ſecond egg. Others have only three hatches, which are regular, conſiſting of three eggs laid without interruption. A fourth ſort, called the *common,* becauſe the moſt numerous, may have four hatches of four or five eggs, but not always uniform. Others are ſtill better layers, having five hatches, and ſtill more if allowed; and in each of theſe are ſix or ſeven eggs. When this kind feed well, they are excellent, and we cannot be too careful of them, for they are worth half a dozen of the ordinary Canaries."

Traité des Serins des Canaries.

of

of plumage is tedious, and commonly more dangerous than in the other kinds. The female Jonquils have only three hatches, each of three eggs; the flaxen coloured Canaries are too delicate, and their brood ſeldom thrives. The Pink ſhew a reluctance to pair with each other; in a large volery, the male but rarely couples with the female of his own colour, and to form the union, they muſt be confined together in a cage. The white ſort are commonly valuable in every reſpect; they lay and breed as well, or rather better than the others, and the variegated white are the hardieſt of all.

Whatever differences there may be in the diſpoſitions and prolific powers of theſe birds, the period of incubation is the ſame; all of them ſit thirteen days; and when the excluſion of the young is a day earlier or later, it is owing to ſome particular circumſtance. Cold retards the proceſs, and heat forwards it; accordingly the firſt hatch, which is in April, requires thirteen days and a half, or even fourteen days, if the weather is chilly; but the third, which happens during the heats of July and Auguſt, is effected in twelve days and a half, or even twelve days. It would be proper to ſeparate all the good eggs from the bad; and, to do this with certainty, we ought to wait to the eighth or ninth day, and take each gently by the two ends for fear of breaking it, and examine it in a ſtrong light, or by a candle, and reject all that are addle, which if left would

would only fatigue the hen. By this trial we may often reduce three hatches to two; and in that caſe the third female may be liberated, and permitted to begin a ſecond neſt *. A plan ſtrongly recommended by bird-fanciers is to remove the eggs as they are dropt, and to ſubſtitute in their ſtead eggs of ivory; ſo that after the laying is over, the real eggs are reſtored, and all hatch at the ſame time. Commonly the egg is dropt at ſix or ſeven o'clock in the morning, and it is ſaid that if it be a ſingle hour later the bird is ſick; and, as the laying proceeds thus regularly, it is eaſy to remove the eggs as faſt as they are excluded †. But this precaution is more ſuited to the convenience of man than conſonant to the train of Nature. When five or ſix young are hatched at once, they exhauſt the vigour of the mother, and rather damp her ſpirits; but if they appear ſucceſſively, they repeatedly renew her pleaſure, and inſpire new courage to diſcharge her duty. Intelligent perſons who have

* When the eggs of one female are ſet under another, they muſt be all ſound. If addle or tainted eggs be given the variegated females, they will throw them out of the neſt; and if the neſt is too deep to admit their being tumbled out, they peck them till broken, which often ſpoils the neſt, and fruſtrates the whole hatch. Females of other colours cover wind eggs when placed under them. *Note communicated by Father* BOUGOT.

† The laying is always at the ſame hour, if the female is healthy; however, the laſt egg muſt be excepted, which is commonly ſeveral hours later, and often a day. This laſt egg is always ſmaller than the reſt; and I am told that it always gives a cock-bird. It would be curious to aſcertain this ſingular fact.

had

had experience in breeding theſe birds aſſure me, that they always ſucceed beſt when they do not employ this artificial expedient.

We may aſſert that, in general, all the nice precautions, and the refined manœuvres recommended by writers for training the Canaries, are pernicious rather than uſeful; and that in every reſpect we ought as much as poſſible to copy Nature. In their native country they haunt the ſides of ſmall rivulets, or wet gullies *; we ought therefore to give them plenty of water, both for drinking and bathing. As they belong to an exceedingly mild climate, they muſt be ſheltered from the rigours of winter. It appears indeed that being long naturalized in France, they can bear the cold of that country; for they may be kept in a chamber without a fire, and even without a glaſs-window, a wire-grate to prevent their eſcape being ſufficient; ſeveral dealers in birds have informed me that they loſt fewer in this way than when the rooms were heated by a fire. The ſame may be ſaid of their food, which is probably the more ſuited to them the ſimpler it is †. A circumſtance that requires the

* The Canary Finches imported into England are bred in the *barancos*, or gullies formed by the torrents from the mountains. *Hiſt. Gen. des Voyages.*

† I have often obſerved from my own experience, and from that of others who adhered ſcrupulouſly to all the minute directions given by authors, that extreme care and attention often killed their birds. A regular diet of rape-ſeed and millet, water every day in winter, and

the moſt attention is not to haſten their firſt hatch; it is common to allow them to couple about the twentieth or twenty-fifth of March, but it would be better to wait till the twelfth or fifteenth of April; for if the ſeaſon be cold, they are apt to contract a diſguſt to each other; and if the females happen to have eggs, they abandon them, at leaſt till the weather grows warm: and thus a whole hatch is loſt in attempting to accelerate the breeding.

The young Canaries differ from the old ones, both in the colour of their plumage, and in ſome other circumſtances. " A young Canary of the ſame year (obſerved on the thirteenth of December 1772*) had its head, its neck, its back, and the quills of the wings blackiſh, except the four firſt quills of the left wing, and the ſix firſt of the right, which were whitiſh; the rump, the coverts of the wings, the tail, which was not entirely formed, and the under part of the body, were alſo of a whitiſh colour; and there were as yet no feathers on the belly from the *ſternum* to the

and once or twice a-day in ſummer; groundſel, when it is to be had, in the month of May; chickweed in the time of moulting, and inſtead of ſugar, bruiſed oats and Turkey wheat, and above all great cleanlineſs, are all that I would recommend. *Small Tract on the Breeding of Canaries, communicated by M. Batteau, Advocate at Dijon.*

N. B. I muſt here correct a ſmall error. All the bird-catchers whom I have conſulted tell me, that we ought to avoid giving them chickweed in the time of moulting, for it is too cooling, and would protract their ſtate of indiſpoſition. The other directions of Batteau ſeem to me well founded.

* Note communicated by *M. Gueneau de Montbeillard.*

anus.

anus. The lower mandible was impreſſed into the upper, which was thick, and ſomewhat incurvated." As the bird grows up, the arrangement, and the ſhades of colour, change; the old ones can be diſtinguiſhed from the young by their ſtrength, their plumage, and their ſong; the tints are deeper, and more lively; their toes are rougher, and incline more on black, if they are of the gray kind; their nails are alſo thicker, and longer than thoſe of the young ones *. The female is often ſo like the male that they cannot at firſt be diſtinguiſhed; however, the colours are always deeper in the male, the head rather thicker and longer, and the temples of a yellow, more inclined to orange; and under the bill there is a ſort of yellow flame which deſcends lower than in the female; its legs are alſo ſtronger; and laſtly, it begins to warble almoſt as ſoon as it is able to feed by itſelf. It is true that ſome females chant at that tender age with almoſt as much ſpirit as the males: but, joining all theſe marks together, we may be able even before the firſt month to decide which are males or females; after that time there is no more uncertainty in that reſpect, for the ſong of the males then begins to betray their ſex.

Every ſudden utterance of ſound is in animals an obvious ſign of paſſion; and as love is of all the inward feelings that which the ofteneſt,

* Amuſemens Innocens, p. 61 & 62.

and

and the moſt forcibly agitates the frame, the ardor is conſtantly marked by the expreſſion of the voice. The birds by their ſong, the bull by his lowing, the horſe by his neighing, the bear by his loud murmur, &c. all announce the working of the ſame deſire. The appetite is much calmer in the female than in the male, and accordingly it is but ſeldom expreſſed by the voice. The chant of the hen Canary is only a feeble tone of tender ſatisfaction, a coy aſſent to the warm applications of her mate, and inſpired by the eloquence of his warble; but when this paſſion is once excited in her veins, it becomes neceſſary to her exiſtence; and if ſhe be parted from her lover, ſhe ſickens and dies.

It ſeldom happens that the Canaries bred in a chamber are indiſpoſed before laying; ſome males only exhauſt their vigour, and fall victims to the ardor of paſſion. If the female becomes ſick while hatching, her eggs muſt be taken from her, and given to another; for though ſhe recovers ſoon, ſhe would not ſit on them again. The firſt ſymptom of bad health, eſpecially in the males, is ſadneſs; as ſoon as they loſe their natural cheerfulneſs, they ought to be put alone in a cage, and ſet in the ſun in the chamber where the female is lodged. If he becomes bloated, we muſt notice if there be a pimple below his tail; when this pimple is ripe and white, the bird itſelf often pierces it with the bill; but if the ſuppuration advances too

 ſlowly,

ſlowly, we may diſcharge it with a large needle, and then fill the wound with ſpittle without ſalt, which would be too ſmarting. Next day the patient ſhould be let looſe in the chamber, and it will be eaſy to perceive by his treatment of the female, and the fondneſs that he ſhews, whether he is cured or not. In this laſt caſe, we muſt take him again, and blow through a ſmall quill ſome white wine under his wings, place him in the ſun, and notice next day the ſtate of his health. If he ſtill continue dejected, and indifferent to his female, his recovery is now almoſt deſperate; we muſt remove him into a ſeparate cage, and give the hen another male like the one ſhe has loſt; or if ſuch cannot be had, we muſt ſeek one of the ſame ſpecies at leaſt. A greater fondneſs commonly ſubſiſts between thoſe that reſemble each other, except in the caſe of the Pink Canaries, which prefer the females of a different colour; but this new male muſt not be a novice in love, and conſequently muſt have already raiſed a hatch. If the female falls ſick, the ſame treatment may be uſed.

The moſt common cauſe of diſtempers is the too great plenty, or richneſs of food. When theſe birds make their neſts in a cage, they often eat to exceſs, or ſelect the nutritious aliments intended for their young; and moſt of them ſicken from repletion, or inflammation. If they be kept in a chamber, this danger is in a great mea-

meaſure removed, their numbers preventing their gluttony. A male which ſits too long, is ſure of being beat by the other males ; and the ſame is the caſe with the females. Theſe quarrels give them exerciſe, and neceſſarily produce temperance ; and for this reaſon chiefly it is that they ſeldom are ever ſick in a chamber during the time of breeding : their infirmities and diſeaſes appear only after they have hatched ; moſt of them have firſt the pimple which I have mentioned, and then they all undergo the moulting. Some ſupport well this metamorphoſis, and ſtill ſing a part of the day ; but moſt of them loſe their voice, and a few languiſh and die. After the females are ſix or ſeven years old, many of them die in changing their plumage ; the males recover better from the attendant ſickneſs, and ſurvive their mates three or four years. Indeed we muſt conſider moulting as the regular proceſs of Nature, rather than as an accidental diſtemper ; and if theſe birds were not reduced by us to captivity, and rendered delicate by our treatment, they would ſuffer only a ſlight indiſpoſition, and would ſpontaneouſly diſcover the proper remedies : but at preſent it is a grievous ſickneſs, often fatal, and which beſides admits of few remedies ; it is however the leſs dangerous the earlier it happens*. The young Canaries drop

* In the time of moulting, a bit of ſteel, and not of iron, ſhould be put in their water, and changed thrice a-week. No other remedies

drop their feathers the firſt year ſix weeks after they are hatched; they become low-ſpirited, appear bloated, and conceal the head in their plumage: at this time the down only falls; but the following year they loſe the quills, even thoſe of the wings, and of the tail. The young birds of the later hatches which happen in September or after, ſuffer much more from the moulting, than thoſe which are excluded in the ſpring; in that delicate condition the cold is extremely pernicious, and they would all periſh if not kept where it is temperate, or even pretty warm. As long as the moulting laſts, that is ſix weeks or two months, Nature labours at the production of new feathers; and the organic molecules, which were before directed to the ſupply of the ſeminal liquor, are now abſorbed in this growth; and hence the exuberance of life being diverted into different channels, their ardor ceaſes, and the buſineſs of propagation is for the time ſuſpended.

The moſt fatal and the moſt common diſtemper, eſpecially in young Canaries, is what is called the *ſwallow* (*avalure*), in which the bowels ſeem to be *ſwallowed*, and drawn to the extremity of their body; the inteſtines are per-

medies are needed, though Hervieux reckons ſeveral; only during this critical period, a rather larger portion of hemp-ſeed ſhould be mixed with their uſual food. *Note communicated by Father* Bougot.

Obſerve that ſteel is recommended inſtead of iron, only leſt the iron ſhould be ruſty, in which caſe it would be more pernicious than uſeful.

ceived

ceived through the ſkin of the belly in the ſtate of inflammation, redneſs, and diſtenſion; the feathers on that part drop, the bird pines, gives over eating, though always ſitting beſide the food, and dies in a few days. The ſource of the diſeaſe is the exceſſive abundance or richneſs of the aliments. All remedies are vain; and the change of regimen is the only thing which can recover a few out of a great number. The bird is put into a ſeparate cage, and given water and lettuce ſeeds; thus the heat that conſumes it is tempered, and evacuations are ſometimes performed which ſave its life. This diſtemper alſo is the fruit of their artificial education, for it ſeldom attacks thoſe which are trained by their parents; we ought therefore to take the greateſt care not to overfeed them with the ſtick; boiled rape-ſeed, and ſome chickweed, are proper, but no ſugar or biſcuit; and in general we ſhould give them too little rather than too much.

When the Canary utters a frequent feeble cry, which ſeems to come from the bottom of its breaſt, it is ſaid to be aſthmatic; it is alſo ſubject to a certain obſtruction of voice, eſpecially after moulting. This ſort of aſthma is cured by giving it the ſeeds of plantain, and hard biſcuit ſoaked in white wine; and to reſtore its voice it ought to have generous food, ſuch as yolks of eggs beat up with crumbs of bread; and for drink, liquorice-water, that is water in which that root has been ſteeped and boiled.

The Canaries are alſo ſubject to a ſort of ſhanker on the bill. This diſorder is owing to the ſame cauſe with the *ſwallow*, the abundance or richneſs of food producing an inflammation, which, inſtead of affecting the inteſtines, ſometimes extends to the throat or palate; the ſame cooling remedies muſt be applied; they ſhould be given lettuce-ſeeds, and bruiſed melon-ſeeds mixed with their drink.

The mites and the ſcab with which theſe ſmall birds are ſometimes affected are generally owing to the dirtineſs with which they are kept. Care muſt be taken to preſerve them clean, to give them water to bathe in, to avoid putting old or bad wood in their cages, and to cover them only with new cloth that is not moth-eaten; and the ſeeds and herbs with which they are to be fed ſhould be fanned and waſhed. We muſt pay this attention, if we would wiſh them to be neat and healthy. In the ſtate of nature they would themſelves preſerve cleanlineſs; but impriſoned, they are ſubject to the loathſome diſorders incident to that ſtate: however, many of theſe birds, though reduced to the melancholy condition of captivity, are never ſick, and in theſe habit ſeems to have become a ſecond nature. In general, the ſource of their diſeaſes is the heat of their conſtitution. They always need water; and if a plate with ſnow be placed in the cage, or even in the volery, they will roll in it ſeveral times with expreſſions of pleaſure,

though

though in the coldeft weather. This proves that it is rather pernicious than ufeful to keep them in very hot places*.

But there is another diftemper to which the Canaries and many other birds are fubject †, efpecially in the ftate of confinement: this is the epilepfy. The yellow Canaries in particular are oftener than the others feized with the falling ficknefs, which attacks them fuddenly, and even furprifes them in the midft of their moft impaffioned warble. It is faid that they muft not be touched the moment they fall, but muft be watched till they difcharge a drop of blood from the bill, and that they may be then lifted up, and will recover themfelves, and in a fhort time refume their fenfes and their life: that it is neceffary to wait till Nature makes that falutary effort which is announced by the expreffion of the drop of blood, and that if handled prematurely, the violence of agitation would bring on inftant death. It is to be wifhed that this obfervation were afcertained, fome circumftances in which appear to me rather doubtful. Certain it is, that if they efcape the firft attack of this

* Thefe birds require not to be kept in a warm place, as many pretend; in the moft intenfe colds they welter in fnow, when prefented them on a plate. For my own part, I have them in a chamber in winter with only an iron-grating, and the windows open; they fing admirably, and I never lofe any. *Note communicated by Father* BOUGOT.

† The Jays, the Goldfinches, all the Parrots, and the largeft Aras.

 epilepfy,

epilepſy, they live a long time after, and ſometimes attain the ſame age with thoſe which have never been affected by that diſtemper. However, I am inclined to think, that a ſmall inciſion in the toes would be beneficial, for in that way Parrots are cured of the epilepſy.

What miſeries in the train of ſlavery! Would theſe birds, if they enjoyed their native freedom, be aſthmatic, ſcabby, and epileptical? Would they be afflicted by inflammations, abſceſſes, and ſhankers? and the moſt melancholy of diſorders, what is produced by the craving of unſatisfied luſt, is it not common to all beings reduced to captivity? In particular, the females, whoſe feelings are ſo nice and ſo tender, are more ſubject to it than the males. It has been obſerved, that after the hen Canary falls ſick in the ſpring before pairing, ſhe ſhrinks, languiſhes, and dies. The amorous paſſion is awakened by the ſinging of the males around her, while ſhe has at the ſame time no opportunity of gratification. The males, though they firſt feel libidinous deſires, and always appear more ardent, ſupport better the languor of celibacy; they ſeldom die of continence, but they are often killed by exceſſive venery.

The hen Canaries can, like the females of other birds, lay eggs without commerce with the male. The egg in itſelf is, as we have elſewhere obſerved, only a matrix which the bird excludes, and will remain unprolific, if not impregnated with the ſeed of the male; and the heat of incubation, inſtead of quickening it, only haſtens its putrefaction. It has alſo been remarked that if the females be entirely ſeparated from the males, ſo as not to ſee and hear them, they very ſeldom lay; and that they ofteneſt drop their eggs, when melted by the view or the ſong of the males: ſo much do even diſtant objects act upon feeling animals, and ſo many are the ways in which the ſubtile flame of love is communicated *!

I cannot better cloſe this article, than by an abſtract of a Letter of the honourable Daines Barrington, Vice Preſident of the Royal Society, to Dr. Maty the Secretary.

" Moſt people who keep Canary birds do
" not know that they ſing chiefly either the
" Titlark or Nightingale notes †.

" Nothing,

* We ſhall here mention two facts to which we were witneſs. A female ſung ſo well, that ſhe was taken for a male, and paired with another female; the overſight being afterwards diſcovered, a male was given to her, who taught her the proper functions of her ſex; ſhe took to laying and renounced her ſong. The other fact is that of a female, alive at preſent, that chants or rather whiſtles a tune, though ſhe has laid two eggs in her cage, which are found to be addle, as all thoſe are which hens lay without the commerce of a cock.

† I once ſaw two or three birds which came from the Canary Iſlands, neither of which had any ſong at all; and I have been informed, that a ſhip brought a great many of them not long ſince, which ſung as little.

Moſt of the Canary birds which are imported from Tyrol, have been educated by parents, the progenitor of which was inſtructed

by

"Nothing, however, can be more marked than
" the note of a Nightingale called its *jug*, which
" moſt Canary birds brought from Tyrol commonly
" have, as well as ſeveral Nightingale
" *ſtrokes*, or particular paſſages in the ſong of
" that bird.

"I mention the ſuperior knowledge in the
" inhabitants of the capital, becauſe I am convinced
" that, if others are conſulted in relation
" to the ſinging of birds, they will miſlead,
" inſtead of giving any material or uſeful information."

by a Nightingale; our Engliſh Canary birds have commonly more of the Titlark notes.

The traffic in theſe birds makes a ſmall article of commerce, as four Tyroleze generally bring over to England 1600 every year; and though they carry them on their backs one thouſand miles, as well as pay twenty pounds duty for ſuch a number, yet upon the whole it anſwers to ſell theſe birds at five ſhillings apiece.

The chief place for breeding Canary birds is Inſpruck and its environs, from whence they are ſent to Conſtantinople, as well as every part of Europe. Phil. Tranſ. vol. lxiii. part 2. 10 January 1773.

FOREIGN BIRDS,

THAT ARE RELATED TO THE CANARIES.

THE foreign birds which may be referred to the Canary, are few in number; we are acquainted with only three ſpecies. The firſt is that which was ſent to us from the eaſtern coaſt of Africa, under the name of *the Mozambique Canary*, which ſeems to be a ſhade between the Canaries and the Siſkins. It is delineated *Pl. Enl.* N° 364, Fig. 1. and 2. Yellow is the prevailing colour of the lower part of the body, and brown that of the upper, except the rump and the coverts of the tail, which are yellow: theſe coverts, as well as thoſe of the wings and their quills, are edged with white or whitiſh. The ſame yellow and brown occur on the head, diſtributed in alternate bars; that which ſtretches over the top of the head is brown, next two yellow ones over the eyes, then two brown ones which riſe behind the eyes, after theſe two yellow ones, and laſt of all two brown ones, which extend from the corners of the bill. This bird is rather ſmaller than thoſe from the Canary Iſlands; its length from the tip of the bill to the end of the tail is about four inches and a half, that of the tail is only about an inch. The female

female differs very little from the male, either in ſize or in colour. This bird is perhaps the ſame with that of Madagaſcar, mentioned by Flaccourt under the name of *Mangoicbe*, which he ſays is a ſpecies of Canary.

It is likely that this bird, which in its plumage reſembles much our variegated Canaries, was their progenitor; and that the entire ſpecies belongs only to the ancient continent, and to the Canary Iſlands, which may be conſidered as adjacent to the mainland: for the one mentioned by Briſſon under the name of *the Jamaica Canary*, and of which Sloane and Ray have given a ſhort deſcription *, appears to me to be widely different from our Canaries, which are not found at all in America. Hiſtorians and travellers inform us, that none were originally in Peru, and that the firſt Canary was introduced there in 1556, and that the ſpreading of theſe birds in

* "Bird like the *Serin*, variegated with cinereous, dilute, yellow, and duſky colours." Its extreme length is eight inches, its alar extent is twelve inches, the bill ſhort and ſtrong, three-fourths of an inch long (or one-third according to Ray), the tail one inch, the leg and foot one inch and one-fourth. (Briſſon ſuſpects that Sloane is miſtaken in his meaſures, for the proportions are not conſiſtent.) The upper mandible brown bordering on blue, the lower lighter coloured; the head and the throat gray; the upper part of the body of a yellow-brown, the wings and the tail of a deep brown, radiated with white, the breaſt and belly yellow, the under part of the tail white, the feet bluiſh, the nails brown, hooked, and very ſhort. SLOANE's *Jamaica*.

It is the *Fringilla Cana* of Linnæus, and *Gray-headed Finch* of Latham: Its ſpecific character:—" It is duſky yellow above, below " yellow, its head and throat gray, its vent white, its wings and " tail duſky, with white lines."

America,

America, and eſpecially in the Antilles, was long poſterior to that date. Father du Tertre relates that du Parquet, in 1657, bought of a merchant who touched at theſe iſlands, a great number of real Canaries, which he ſet at liberty; ſince which time they were heard warbling about his houſe; ſo that it is probable that they have multiplied in that country. If true Canaries be found in Jamaica, they may have been deſcended from thoſe tranſported to the Antilles, and naturalized there in the year 1657. However, the bird deſcribed by Sloane, Ray, and Briſſon, by the appellation of *Jamaica Canary*, appears to differ too much from the natives of the Fortunate iſlands, to be ſuppoſed to have originated from thoſe tranſplanted into the Antilles.

While this article was at the preſs, we received ſeveral Canaries from the Cape of Good Hope, among which I have perceived three males, one female, and a young one of the ſame year. Theſe males are very like the Green Canary of Provence; they differ in being ſomewhat larger, and their bills being proportionally thicker: their wings are alſo better variegated, the quills of the tail edged with a diſtinct yellow, and they have no yellow on the rump.

In the young Canary, the colours were ſtill fainter and leſs marked than in the female.

But whatever ſmall differences exiſt, I am ſtill the more confirmed that the variegated Canaries

Canaries of the Cape of Mozambique*, of Provence, and of Italy, are all derived from the ſame common ſource, and that they belong to one ſpecies, which is ſpread, and ſettled in all the climates of the ancient continent ſuited to its conſtitution, from Provence and Italy to the Cape of Good Hope, and the adjacent iſlands. Only this bird has aſſumed more of the green tint in Provence, more of the gray in Italy, more of the brown or variegated colour at the Cape of Good Hope, and ſeems by the variety of its plumage to point at the influence of a different climate.

* It appears that the Mozambique Canary is not confined excluſively to that region. I have found among the drawings of Commerſon a coloured figure of this bird very diſtinctly marked. Commerſon calls it the Cape Canary, and informs us that it had been carried to the Iſle of France, where it was naturalized, and even greatly multiplied, and was known there by the name of *the Bird of the Cape*. We may expect to find in the ſame manner at Mozambique, and in ſome other countries of Africa, the variegated Canaries of the Cape, perhaps even thoſe from the Fortunate Iſlands, and probably many other varieties of this ſpecies.

The WORABEE.

The ſecond ſpecies which appears to us to be the neareſt related to the Canaries, is a ſmall Abyſſinian bird †, of which we have ſeen

† This is the *Fringilla Abyſſinica* of Gmelin, and the *Black-llared Finch* of Latham. Specific character:—“ It is black, yellow above, its collar black, its vent yellow. ’

the

the figure excellently delineated and coloured by Mr. Bruce, under the appellation of the Worabée of Abyſſinia.

This ſmall bird preſents not only the colours of certain varieties of the Canary, the yellow and the black, but it has the ſame bulk nearly; and except its being rather rounder, the ſame ſhape. Its bill is alſo ſimilar, and it prefers an oily ſeed as the Canary does millet and panic. But the Warabée has an excluſive predilection for a plant that bears the oily ſeed which I have mentioned, and which is called *Nuk** in the Abyſſinian language; it never wanders far from that plant, and ſeldom loſes ſight of it.

In the Warabée, the ſides of the head, as far as below the eyes, the throat, the fore-part of the neck, the breaſt, and the top of the belly to the legs, are black; the upper-part of the head and all the body and the lower belly, are yellow, except a kind of black collar, which encircles the neck behind, and is ſet off by the yellow. The coverts and the quills of the wings are black, edged with a lighter colour; the feathers of the tail are alſo black, but with a greeniſh yellow border; the bill likewiſe is black, and the legs of a light brown. This bird keeps in

* The flower of this plant is yellow, and of the ſhape of a creſcent; the ſtalk riſes only two or three inches. From the ſeed an oil is extracted, which is much uſed by the monks of that country.

flocks,

flocks, but we know nothing more in regard to its mode of life.

The ULTRA-MARINE.

L' Outre-Mer. * Buff.

The third ſpecies alſo of theſe Foreign Birds, which are related to the Canary, is known to us from the drawings of Mr. Bruce. I call this Abyſſinian bird *the Ultra-Marine*, becauſe its plumage is of a fine deep blue. In the firſt year this beautiful colour does not exiſt, and the plumage is gray as that of the Sky-lark, and this gray always continues in the female, but the males aſſume the charming blue the ſecond year, before the vernal equinox.

Theſe birds have a white bill and red legs. They are common in Abyſſinia, and never change their habitation. They are nearly of the bulk of the Canaries, but their head is round; their wings extend beyond the middle of the tail. Their warble is very pleaſant, and this circumſtance ſeems the more to point out their relation to our Canaries.

* The *Fringilla Ultramarina* of Gmelin, and the *Ultramarine Finch* of Latham. Specific character :—" It is cœrulean, its bill " white, its legs red."

The

The HABESH of SYRIA*.

MR. Bruce ſuppoſes this bird to be a ſpecies of Linnet, and I ought to pay deference to the opinion of ſo good an obſerver; but that gentleman having figured it with a thick ſhort bill, very like that of the Canaries, I have ventured to place it between the Canaries and Linnets.

The upper-part of its head is of a fine bright red; the cheeks, the throat, and the upper-part of the neck are a mottled blackiſh brown; the reſt of the neck, the breaſt, the upper-part of the body, and the ſmall coverts of the wings, variegated with brown, yellow, and blackiſh; the great coverts of the wings of a deep aſh, edged with a lighter colour; the quills of the tail and the great quills of the wings alſo cinereous, bordered on the outſide with a bright orange; the belly and the under-part of the tail, dirty white, with obſcure yellowiſh and blackiſh ſpots; the bill and legs of a leaden colour. The wings reach as far as the middle of the tail, which is forked.

* The *Fringilla Syriaca* of Gmelin, and the *Tripoline Finch* of Latham.

The Habesh is thicker than our Linnet; its body is also fuller, and it sings prettily. It is a bird of passage; but Mr. Bruce cannot trace its route, and he assures me, that in the course of his travels he never saw it except at Tripoli in Syria.

No. 88

THE LINNET

The LINNET.

La Linotte, Buff.

NATURE herſelf ſeems to have aſſigned theſe birds a place next after the Canaries; for their mutual commerce ſucceeds better than the intercourſe of either with any other contiguous ſpecies; and what points out the cloſeneſs of this relation, the progeny is prolific *, eſpecially when a male Linnet is joined with a female Canary.

Few birds are ſo common as the Linnet, but ſtill fewer perhaps unite ſo many amiable qualities: a pleaſant warble, a rich plumage, docility of diſpoſition, ſuſceptibility of attachment; poſſeſſing whatever, in ſhort, could invite the attention of man, and contribute to his delight. Endowed with ſuch talents, it could not long preſerve its freedom; and ſtill leſs, when nurſed in the boſom of ſlavery, could it retain un-

* This obſervation was communicated by Daubenton the younger. Friſch aſſures us, that pairing a Vine Linnet (Redpole) with a white hen Canary, that was in the habit of coming abroad every day and returning to its rooſt, it made its neſt and laid its eggs in a neighbouring buſh, and when the young were hatched, it brought them to the window of the houſe. He adds, that this croſs-breed had the whole plumage of the mother, and the red ſpots of the father, eſpecially on the head.

 ſullied

ſullied the beauties of its original purity. In fact, the charming red colour with which Nature has painted its head and breaſt, and which in the ſtate of liberty ſparkles with durable luſtre, wears off by degrees, and ſoon diſappears entirely in our cages and voleries. There remain only a few obſcure veſtiges after the firſt moulting*.

With regard to the change effected in its ſong, we ſubſtitute for the free and varied modulations which ſpring and love inſpire, the ſtrained notes of a harſh muſic, which they repeat but imperfectly, and which has neither the beauties of art nor the charms of nature. Some have alſo ſucceeded in teaching it to ſpeak different languages, that is to whiſtle ſome Italian, French, and Engliſh words, &c. and ſometimes even to pronounce theſe with conſiderable fluency. Many perſons have from curioſity gone from London to Kenſington merely to hear an apothecary's Linnet, which articulated the words *pretty boy*. The fact is, it had been taken out of the neſt when only two or three days old, before it had time to acquire the parent ſong; and juſt beginning to liſten with attention, it was ſtruck with the ſound of *pretty boy*, and learned it from

* The red of the head changes into a ruſty-brown varied with blackiſh, and that of the breaſt paſſes into nearly the ſame colour; but the new ſhades are not ſo deep. An *amateur* told me, that he has raiſed ſome of theſe Linnets which preſerved the red: this fact ſtands ſingle.

imitation.

imitation *. This fact, together with many others, appears to me to establish the opinion of the Honourable Daines Barrington, that birds have no innate song; and that the warble peculiar to the different species, and its varieties, have nearly the same origin with the languages and the dialects of various nations †. Mr. Barrington tells us, that in experiments of this kind he preferred a young cock Linnet of three weeks be-

* A goldfinch which was taken from the nest two or three days after hatching, and set in a window that looked into a garden where the wrens resorted, caught their song, and had not a single note of its own species.

A sparrow was taken from the nest when it was fledged, and educated under a Linnet; but hearing by accident a goldfinch, its song was a mixture of that of the Linnet and the goldfinch.

A robin was set under a very fine nightingale, which began to be out of song, and in a fortnight was perfectly mute; the robin had three-parts in four of the nightingale's warble, the rest being a confused jumble.

Lastly, Mr. Barrington adds, that the Canaries imported from Tyrol seem to have been educated by parents the progenitor of which was instructed by a nightingale; while the Canaries bred in England appear to have derived their song from the tit-lark. *Philosoph. Transact. Jan.* 10, 1773.

If we breed a young Linnet with a chaffinch, or nightingale, says Gesner, it will acquire their song; and particularly that part of the chaffinch's song that is called the *alarum sound*. REITERZU, p. 591.

† The loss of the parent-cock at the critical time for instruction occasions undoubtedly the varieties in the song of each species; because then the nestling has either attended to the song of some other birds, or perhaps invented some notes of its own, which are perpetuated from generation to generation, till similar accidents produce other alterations. The truth is, that scarcely any two birds of the same species have exactly the same notes, if they are accurately attended to, though there is a general resemblance. BARRINGTON, *Philos. Trans.* 1773.

 ginning

ginning to fly, not only on account of its great facility and talent for imitation, but because in that species it is easier to distinguish the sex; some of the outer-quills of the wings in the male having the outer-edge white as far as the shaft, while in the female these are only bordered with that colour.

It follows from the experiments of this learned gentleman, that the young Linnets educated by the different kinds of larks, and even by an African Linnet, called *Vengolina*, of which we shall afterwards treat, acquired not the song of their parent, but that of their instructor. Only a few retained the *call* of its species, or the Linnets *chuckle*, which they had heard of their parents before they were parted.

It is extremely doubtful whether our Common Linnet, called by some the *Gray Linnet*, is different from that termed the *Vine Linnet*, or the *Red Linnet*: for, 1. The red spots which distinguish the males in the Red Linnet are far from being constant, since, as we have already observed, they become obliterated in confinement*. 2. They are not a discriminating character,

* Of four cock Linnets, which were consequently red, brought to me on the twelfth of July, I exposed one to the open air, and set three in a chamber, two of which were shut up in the same cage. The red on the head of the latter began to disappear by the twenty eighth of August, and also that on the lower part of the breast. On the eighth of September one of the two was found dead; its head was entirely divested of feathers, and even slightly wounded. I discovered that the one had fought the other after the moulting, as if their

racter, as traces of them are to be found in the bird deſcribed to be the male of the Gray Linnet, where the feathers on the breaſt are of a dull red in their middle. 3. The moulting tarniſhes, and for a time almoſt diſcharges this red, which recovers not its luſtre till the fine weather, but from the end of September colours the middle of the feathers on the breaſt, as in that reckoned by Briſſon a common cock Linnet. 4. Geſner at Turin, Olina at Rome, Linnæus at Stockholm *, and Belon in France, have known in their reſpective countries only the Red Linnets. 5. Bird-catchers, who have in France followed that profeſſion more than thirty years, have never found a ſingle cock Linnet which had not the red ſhade correſponding to the ſeaſon; and at the ſame time we ſee many Gray Linnets in the cage. 6. Even thoſe who admit

their acquaintance was diſſolved by the diſguiſe of plumage. The red of the head of the vanquiſhed Linnet no longer exiſted, for all the feathers had dropped, and that of the breaſt was more than half effaced.

The third which was ſhut up was very late in moulting, and retained its red till that time. The one that was kept in the open air made its eſcape at the end of three months; but it had already loſt all its red.—It follows from this experiment, that either the open air haſtens the diſappearance of the red, by advancing the moulting; or that the want of freſh air has a ſmaller ſhare in the change of plumage in theſe birds than the loſs of liberty.

* No mention is made of the Gray Linnet in the *Fauna Suecica*. Klein ſpeaks of one Zarn, author of a letter on the birds of Germany, where he endeavours to prove that there is only one ſpecies of Linnet. I have heard the ſame aſſertion of many bird-catchers, who had never ſeen the letter; and Hebert, who is certainly a fit judge of the matter, is of the ſame opinion.

the exiſtence of Gray Linnets in the ſtate of nature agree, that they are ſcarcely ever caught, particularly in ſummer, which they attribute to their ſhy diſpoſition. 7. Add to all theſe, that the Red and Gray Linnets are very ſimilar in the reſt of their plumage, in their ſize, in the proportions and ſhape of their parts, in their ſong, and in their habits. And it will be eaſy to infer, that if Gray Linnets really do exiſt, they are either, 1. all females; or, 2. all young males of the year's hatch before October, for at that time they begin to be marked; 3. ſuch as being bred apart from the mothers cannot aſſume red in the ſtate of captivity; 4. thoſe which being caught loſe their tint in the cage*; or, laſtly, thoſe in which this beautiful colour is effaced by moulting, diſeaſe, or ſome other cauſe.

The reader will not then be ſurpriſed that I refer theſe two Linnets to the ſame identical ſpecies; and conſider the gray ſort as only an accidental variety, partly effected by education, and afterwards miſtaken by authors.

The Linnet often builds its neſt in vineyards, and hence it has been called the *Vine-Linnet*. Sometimes it places its neſt on the ground; at other times it fixes it between two props, or

* We muſt obſerve that thoſe birds which have had the red ſpots, but loſt them, ſtill retain on the ſame parts a rufous colour, approaching to red; which never appears in the young that have been bred without the mothers, and that conſequently have never been marked with red.

even

even in the vine itſelf; it breeds alſo in juniper, gooſeberry, and hazel-nut trees, in young copſes, &c. A great number of theſe neſts have been brought to me in the month of May, a few in July, and only one in September: they were all compoſed of ſlender roots, ſmall leaves, and moſs on the outſide, and lined with feathers, hairs, and a great deal of wool. I never found more than ſix eggs; that of the fourth of September had only three; they were of a dirty white, ſpotted with brown red at the large end. The Linnets have ſeldom more than two hatches, except their eggs be robbed, which obliges them to renew their labours; and in this way they may be made to lay four times in the year. The mother feeds the young by diſgorging into their bill what had been prepared and half-digeſted in her craw.

After the hatching is over, and the family raiſed, the Linnets go in numerous flocks, which are formed about the end of Auguſt, when the hemp is arrived at maturity; and at this time ſixty have been caught in one drawing of the net*, and out of theſe were forty males. They continue to live thus in ſociety during the whole winter; they fly very crowded, alight and riſe together, perch on the ſame trees, and about the

* The lark-net may be uſed, but it ought to be rather ſmaller and cloſer. One or two cock Linnets ſhould be had for calls. Chaffinches, and other ſmall birds, are often caught with the Linnets.

begin-

beginning of ſpring they all chant at once; they lodge during the night in oaks, and elms, whoſe leaves, though dry, have not yet fallen; they are ſeen too on the linden-trees and poplars, and feed upon the buds; they live alſo upon all kinds of ſmall ſeeds, particularly thoſe of thiſtles, &c. and hence they haunt indiſcriminately uncultivated lands, and ploughed fields. Their walk is a ſort of hopping; but their flight is continued and uniform, and not like that of the ſparrow, compoſed of a ſucceſſion of jerks.

The ſong of the Linnet is announced by a ſort of prelude. In Italy the Linnets of Abruzzo and of the Marche of Ancona are preferred. It is generally ſuppoſed in France that the warble of the Red Linnet is ſuperior to that of the Gray. This is conſonant to reaſon; for a bird which has formed its ſong in the boſom of liberty, and from the impulſe of its inward feelings, muſt have more affecting and expreſſive airs, than one that has no object but only to cheer its languor, or to give the neceſſary exerciſe to its vocal organs.

The females are naturally deſtitute of ſong, nor can they ever acquire it. The adult males caught in the net profit as little by inſtruction; and the young males taken out of the neſt are alone ſuſceptible of education. They are fed with oatmeal-gruel, and rape-ſeed ground with milk or ſugared-water; and in the evening they are whiſtled to in the weak light of a candle, care being taken to articulate diſtinctly the words which they are wanted to repeat. Sometimes, to begin them, they are held on the finger before a mirror, in which they view their image, and believe they ſee another bird of their own ſpecies; they ſoon fancy that they hear the notes of a companion, and this illuſion produces a ſort of emulation which animates their ſongs, and quickens their progreſs. It is ſuppoſed that they ſing more in a ſmall cage than in a large one.

The very name of theſe birds points out their proper food. They are called Linnets (*Linariæ*) becauſe they prefer lint-ſeed; to this are added the ſeeds of panic, of rape, of hemp, of millet, of Canary-graſs, of raddiſh, of cabbage, of poppy *, of plantain, of beet, and ſometimes thoſe of melon bruiſed. From time to time they have ſweet cake, prickly-ſorrel, chickweed, ſome ears of wheat, oats pounded, and even a little ſalt; but all theſe muſt be properly varied. They break the ſmall grains in their bill, and reject the ſhell; they ought to have very little hemp-ſeed, for it fattens them too much; and this exceſſive fat occaſions their death, or at leaſt renders them unfit for ſinging. In thus feeding and raiſing them one's ſelf, we ſhall not only teach them what airs we chuſe with a Canary-whiſtle, a flageolet, &c. but we ſhall tame them.

* Geſner ſays that if poppy-ſeeds alone be given for food either to Linnets or Goldfinches, they will become blind.

Olina

Olina adviſes to ſhelter them from cold, and even to employ remedies for their diſeaſes; that we ought for inſtance to put in their cage little bits of plaſter to prevent coſtiveneſs, to which they are ſubject: he directs oxymel, ſuccory, &c. in caſes of aſthma, phthiſic*, and certain convulſions, or beatings with the bill. This laſt, however, I ſhould ſuppoſe, is only a kind of careſſing; the little animal, overcome by inward workings, makes the moſt violent efforts to communicate its ſentiments. At any rate we muſt attend much to the choice and quality of the grain that is given it, and obſerve great cleanlineſs in its food, crink, and volery; when ſuch care is taken, the bird may live in confinement five or ſix years, according to Olina; and much longer according to others †. They diſtinguiſh thoſe who are kind to them, become fond of them, alight on them out of preference, and behold them with an affectionate air. If we would abuſe their docility, we might even make them draw water; for they acquire habits as readily as the Siſkin and Goldfinch. They begin to moult about the dog-days, and ſometimes much later: a Linnet and a Siſkin have been known not to drop their feathers be-

* The captive Linnets are alſo ſubject to the epilepſy, and the boil. Some ſay that they can ſcarcely ever be cured of this boil; others direct to puncture it ſeaſonably, and pour wine into the wound.

† There is one at Montbeillard that is certainly known to be ſeventeen years old.

fore October; they had sung till that time, and their music was superior to that of any other bird in the same volery; and their moult, though late in the season, was expeditious and easy.

The Linnet is a pulverulent bird, and it would be proper to strew in the bottom of the cage a layer of fine sand, and renew it occasionally; there ought also to be a small bath.—The total length of the bird is five inches and a few lines; its alar extent nearly nine inches; its bill five lines; its tail two inches, somewhat forked, and stretching an inch beyond the wings.

In the male the top of the head and the breast are red; the throat, and the under-part of the body, rusty-white; the upper-part, chesnut; almost all the feathers of the tail and of the wings are black, edged with white; and hence when the wings are closed there is a white ray parallel to the feathers. The female has commonly none of the red that we have mentioned; and the plumage is more varied than in the male. [A]

[A] The two kinds of Linnets which Buffon conceives to be originally the same, are distinguished by systematic writers:—

1. *The Common Linnet*, called sometimes the *Gray Linnet*. It is the *Fringilla Linota* of Gmelin; the *Linaria* of Ray, Willughby, Brisson, Frisch, &c. The German name is *Flacks-Finch*, that is *Flax-Finch*; the Italian, *Fanello*; the Dutch, *Knue*; the Brabantish, *Vlasvinch*; the Turkish, *Gezegen*. Aristotle termed it Αιγιθος. The character given by Brisson is:—" It is of a dusky chesnut, and " beneath tawny white; its wings are marked by a longitudinal " white stripe; its tail-quills are black, edged with white." In the beginning of the spring the breast of the male is of a rose-crimson

ſon colour, which does not take place in the female. It builds its neſt with moſs and bents, and lines it with wool and hair; lays five eggs.

2. *The Greater Red-headed Linnet*, or *Red-Poll*, which Buffon terms *The Greater Vine-Linnet*. It is the *Fringilla Cannabina* of Linnæus, or the *Hemp-Finch*; the *Linaria Rubra* of Geſner, Ray, Briſſon, &c. The German name is *Hänfling*, or *Hemp-bird*; and the Italian *Fanello Marino*, or *Sea-Linnet*. It is thus characterized by Briſſon:—"It is of a duſky-cheſnut, the margins of its fea-"thers more dilute, and beneath of a tawny-white; its wings "marked with a white longitudinal ſtripe; its tail-quills black, "the whole of their borders white." The Linnæan character:—"The primary quills of its wings, and thoſe of the tail, black, "and white at both the edges." It is found both in Europe and in America. It is ſmaller than the preceding, and is gregarious in winter. The female has neither the red ſpot on the crown, nor the bluſh-coloured breaſt. It neſtles on the ground. It is a very familiar bird, and quite cheerful a few minutes after it is caught.

The opinion of our ingenious author with reſpect to the identity of the ſpecies of the Red Poll, and of the Common Linnet, is very plauſible; but it ſeems not altogether well founded. The Red Poll is ſmaller than the Linnet; it neſtles on the ground, while the latter breeds in furze and thorn-hedges. The egg of the Linnet is of a very faint blue, dotted with ruſty ſpecks, and interſperſed with minute brown ſtreaks. The egg of the Red Poll is a very faint green, ſprinkled with ruſty dots, and rather ſharp at the ſmall end.

VARIETIES of the LINNET.

I. The White Linnet. I have seen this variety at the house of Desmoulin, the painter. White was the predominant colour of its plumage, but the quills of the wings and of the tail were black, edged with white, as the Common Linnet, and some vestiges of gray also were perceptible on the upper coverts of the wings.

II. The Black-legged Linnet. Its bill is greenish, and the tail much forked; in other respects, it is the same in size, in proportions, and even in colours, with the common Linnet. This bird is found in Lorraine, and we are indebted for our information to Dr. Lottinger of Sarbourg.

M

The STRASBURG FINCH.

Le Gyntel de Strasbourg, Buff.
Fringilla Argentoratensis, Gmelin.
Linaria Argentoratensis, Briss.

Little is known with regard to this bird, yet enough to indicate its affinity to the Linnet. It is is of the ſame ſize, it feeds upon the ſame ſeeds, it flies alſo in numerous flocks, and has eggs of the ſame colour: its tail is forked, the upper-part of its body of a deep brown, the breaſt rufous, ſpeckled with brown, and the belly white. It lays indeed three or four eggs only, according to Geſner, and its legs are red. But was Geſner accurately acquainted with the number of eggs? and with regard to the red colour of the legs, we have ſeen, and we ſhall have other opportunities of being convinced, that this property is far from being foreign to Linnets, eſpecially to thoſe in their natural ſtate. The analogy appears even amidſt the differences, and I am inclined to believe, that when the Straſburg Finch is better known, it may be referred as a variety derived from climate, ſituation, &c. to the common Linnet *.

* Specific character:—" It is duſky, above rufous, ſpotted " with duſky, its lower belly and vent whitiſh." It is thus deſcribed by Briſſon; " above it is duſky, below rufous, variegated " with duſky ſpots, its lower belly whitiſh, its tail-quills duſky, its " legs reddiſh."

The MOUNTAIN LINNET.

La Linotte de Montagne, Buff.
Fringilla Montium, Gmel.
Linaria Montana, Briss.

This bird is found in the mountainous part of Derbyshire in England *: it is larger than the ordinary sort, and its bill proportionally more slender. The red, which appears on the head and breast of the common cock Linnet, occurs in the male of this species on the rump. In other respects the plumage is nearly the same. The breast and throat are variegated with black and white; the head with black and cinereous, and the back with black and rusty. The wings have a transverse white ray, which is very distinct, being on a black ground; it is formed by the great coverts which are tipt with white. The tail is two inches and a half long, composed of twelve brown quills, of which the lateral ones have a white edging, which is broader the nearer the quill lies to the outside.

It is probable that the Mountain Linnet has a forked tail, and that its warble is pleasant, though

* May not this be the second Linnet mentioned by Gesner, and after him by Schwenckfeld, which is more shy than the common Linnet, has an inferior song, and inhabits arid mountains, at least if we judge from the name that he gives it, *Stein Haënfling* (*Stone Linnet*)?

 Willughby

Willughby does not exprefsly mention thefe circumftances. But he reckons thefe two characters peculiar to the Linnet, with which he ranges it. If we be permitted to draw this inference, we might confider the Mountain Linnet as only a variety refulting from climate or fituation. [A]

[A] The Mountain Linnet advances during the winter in flocks to the fouthern parts of England, and perhaps breeds in the northern counties. It is larger than the Red Poll, being fix inches and a half long. Specific character:—" It is black, below pale, " the throat and breaft blackifh, the rump in the male pale-red." Thus defcribed by Briffon:—" Above, it is black and variegated " with tawny, below whitifh; the feathers of the lower part of " the neck and of the middle of the breaft are black, (the rump " red in the male,) with a white tranfverfe ftripe; the quills of the " tail dufky, the edges of the lateral ones white on both edges." Its egg refembles that of the common Linnet in colour, but the fpecks are thinner fcattered, and its fhape is rather more bulged.

M

The TWITE.

Le Cabaret, Buff.
Fringilla Montium, Var. Linn.
Linaria Minima, Briff.
Linaria Pectore Subluteo, Klein.
Fanello dell' Aquila, Olin.

When we treat of birds whofe colours are fo variable as in the prefent, we fhould fall into numberlefs miftakes, if we confidered them as fpecific characters. We have already feen that the

the common Linnet, in the ſtate of liberty, was marked with red on the head and breaſt; that the captive Linnet had the ſame colour only on the breaſt, and that concealed; that the Straſburg Linnet had it on the legs; and that the Mountain Linnet was painted with it on the rump. Briſſon ſays, that what he calls the Little Vine Linnet is red on the head and breaſt; and Geſner adds, on the rump. Willughby mentions a ſmall Linnet which had a red ſpot on the head, and in that circumſtance reſembled the two deſcribed by Aldrovandus, though it differed in other reſpects. Laſtly, the *Cabaret* of Briſſon was marked with red on the head and rump, but that of Friſch had none on the head. It is obvious, that a great part of theſe varieties is owing to the ſeaſons and the circumſtances in which theſe birds have been ſeen. If in the middle of ſpring, they were clothed in their richeſt colours; if during the moulting ſeaſon, the red had diſappeared; if immediately after, it was not yet reſumed. If they were kept in a cage, the change would be in proportion to the length of their confinement; and as the feathers of the different parts of the body would drop at different times, there would be abundant ſource of diverſity. In this uncertainty, we are obliged, in order to determine the ſpecies, to recur to the more permanent properties; the ſhape of the body, the inſtincts, and habits. Applying this method, I can perceive only two kinds of birds that have received the name of Little Linnet: the one, which never ſings, appears but once in ſix or ſeven years, arrives in numerous flocks, reſembles the Siſkin, &c.: it is the Little Vine Linnet of Briſſon: the other is the Twite of this article.

The younger Daubenton had for two or three years one of theſe birds, which was caught in a net. It was at firſt very ſhy, but it became gradually reconciled, and grew quite familiar. It ſeemed to prefer hemp-ſeed for its food. It had a ſweet mellow note, much like that of the Hedge Pettychaps. It loſt all its red the firſt year, and never recovered it; the other colours ſuffered no alteration. When ſick or in moult, its bill was obſerved to become immediately pale and yellowiſh; and as the bird recovered, it paſſed through all the ſhades to its proper brown caſt. The female is not entirely deſtitute of brilliancy of plumage; it is marked with red on the head, but not on the rump. Though ſmaller than the common hen Linnet, its voice is ſtronger and more varied. This bird is rare both in Germany and in France. It flies rapidly, but never in large flocks; its bill is rather more ſlender in proportion than that of the common Linnet.

Meaſures: the total length of the Twite is four inches and a half; its wings are eight inches acroſs; its bill rather more than four lines; its tail two inches: this is forked, and projects only eight lines beyond the wings.

Colours:

Colours: The upper-part of the head and the rump are red; there is a reddiſh bar under the eyes; the upper-part of the body is variegated with black and ruſt colour; the under-part of the body ruſty, ſpotted with blackiſh below the throat; the belly is white; the legs brown, ſometimes black. The nails are very long, and that of the hind toe is longer than the toe itſelf. [A]

[A] Linnæus makes the Twite a variety of the Mountain Linnet; but Mr. Latham conceives it to be more related to the Red Poll. "It is variegated above, rufous below, the abdomen "whitiſh, the eye brows and the bar on the wings tawny, the "crown and rump red." It is about four inches and a half long. Briſſon deſcribes it "as blackiſh above and varied with rufous, "below tawny; the belly whitiſh; the ſtripe above the eyes ru-"fous; having ſpots between the bill and the eyes and under the "throat of a duſky blackiſh colour, the crown and rump red in the "male; (the crown red in the female;) the tranſverſe bar on "the wings of a white-rufous; the tail quills duſky, the whole "of the edges tawny." If we were to judge from the egg of the Twite, we ſhould regard it as a variety of the Red Poll. The egg of the Twite, though rather ſmaller than that of the Red Poll, is of the ſame ſhape and colour; only the dots are orange, ſmaller, and more thickly ſpread.

FOREIGN BIRDS, THAT ARE RELATED TO THE LINNET.

I.

The VENGOLINE*.

ALL that is known with regard to the history of this bird is, that it is found in the kingdom of Angola; that it is very familiar; that it is ranked among the finest warblers of that country; and that its song is not the same with that of our Linnet. The neck, the upper-part of the head and of the body, are variegated with two sorts of brown; the rump has a beautiful spot of yellow,

* This name was applied to it by Mr. Daines Barrington. It is the *Angola Finch* of Latham, the *Fringilla Angolensis* of Gmelin, and the *Linaria Angolensis* of Brisson. " It is dusky-cinereous spotted " with dusky, below orange, the throat and the fore-head black, " the cheeks and throat spotted with white, the rump yellowish." Brisson describes it: MALE;—" Tawny-cinereous above, variegated with dusky spots; scarlet below; the throat lighter coloured; the small feathers round the base of the bill and on the " throat black; the cheeks and throat spotted with white; the " rump yellowish; the tail feathers dusky, edged with cinereous " white." FEMALE;—" Tawny-cinereous above, rufous with " dusky spots below, and variegated both above and below; a " dusky stripe stretching on either side over the eyes; the cheeks " light tawny; the rump whitish; the tail feathers dusky, edged " with cinereous white."

 which

which extends to the quills of the tail: theſe quills are brown, edged and tipt with light gray, as alſo the quills of the wings, and their great and middle coverts. The ſides of the head are of a light ruſt colour, and there is a brown ſtreak below the eyes; the under-part of the body and the ſides are ſpotted with brown on a lighter ground.

Edwards, who firſt deſcribed this bird, and who has given a figure of it at the bottom of Pl. 129, is inclined to think it is a female of another that is repreſented at the top of the ſame plate. Thisother bird is called *Negral* or *Tabaque*, and its ſong is much like that of the Vengoline. For my own part, I muſt confeſs, that the ſong * of this makes me doubt of its being a female. I ſhould rather ſuppoſe that they are two males of the ſame ſpecies, but from different climates, where each would have a diſtinct name; or at leaſt that they are two males of the ſame climate, one of which being bred in a volery, had loſt the luſtre of its plumage, and the other not being caught before it was adult, or having been kept but a ſhort time in the cage, had preſerved its colours better. In fact the colours of the Negral are richer and more marked than thoſe of the Vengoline. The throat, the face, and the ſtreak below the eyes, are black;

* Daines Barrington ſays, that the Vengoline excels in point of ſong all the birds of Aſia, Africa, and America, except only the American Mocking Bird.

 the

the cheeks white, the breaſt and all the under-part of the body of an orange colour, without ſpecks, and which aſſume a deeper ſhade under the belly and the tail. Theſe two birds are of the ſize of our Linnet; and Edwards adds, that they have the ſame aſpect.

M

II.

The GRAYFINCH*.

La Linotte Gris-de-Fer, Buff.
Loxia Cana, Linn.
Fringilla Cinerea Indica, Briſſ.
Cinereous Groſbeak, Lath.

We are indebted to Edwards for the knowledge of this bird, who had it alive, and has given a figure and deſcription of it, without informing us from what country it came. Its warble is very pleaſant. It has the geſtures, the ſize, the ſhape, and the proportions of the Linnet, only its bill is rather ſtronger. The under-part of its body is of a very light aſh-colour, the rump rather darker; the back, the neck, and the upper-part of the head iron-gray; the quills of the tail and of the wings blackiſh, edged with

* Specific character:—"Hoary, the feathers of the wings and "of the tail duſky, the legs red."

light

light cinereous, except the longeſt quills of the wings, theſe being entirely black near the end, and white at their origin; which gives the wings a white border in its middle-part. The lower mandible is encircled at its baſe with white, and this colour extends as far as the eyes.

M

III.

The YELLOW-HEADED LINNET.

La Linotte à tête jaune, Buff.
Loxia Griſea, Gmel.
Paſſer Mexicanus, Briſſ.
Emberiza Flava Mexicana, Klein.

Edwards knew that this bird was called by ſome *the Mexican Sparrow**, and he has ranged it with the Linnets, becauſe it is more related

* Specific character:—" Cœrulean gray, face and lower-part of " the neck white." Thus deſcribed by Briſſon:—" Above dirty " coloured, varied with black ſpots, below ſlightly duſky with " dull ſpots, and variegated with duſky ſpots; the fore-part of " the head, the cheeks and the throat yellowiſh; a duſky longi- " tudinal bar behind the eyes; the tail-quills blackiſh."

Dr. Fermen, in his deſcription of Surinam, mentions a *Linnet with a yellow throat and bill*, the reſt of the plumage being cinereous. " It haunts the Savannas, and is larger than a Spar- " row." . . Its ſong is not ſo pleaſant as to deſerve to be kept in a cage, but it is reckoned as a ſort of Ortolan, becauſe it is delicate eating.

to

to theſe than to the Sparrows. It is indeed true, that it alſo bears ſome analogy to the Canaries, and we might place it with the Habeſh, between the Linnets and Canaries; but the obſcurity of its hiſtory renders it more difficult to aſſign it the proper rank.

It is of a pale fleſh colour, the feet of the ſame, but duſkier; the fore-part of the head and throat yellow, and upon this yellow ground there is a brown bar on each ſide of the head, riſing from the eye and deſcending on the ſides of the neck; all the upper-part of the body is brown, but deeper on the quills of the tail than any where elſe, and ſprinkled with lighter ſpots on the neck and back; the lower-part of the body is yellowiſh, with brown longitudinal ſpots ſcattered thinly over the belly and breaſt.

This bird was brought from Mexico. Briſſon ſays that it is nearly of the ſize of the Brambling; but if we judge by the figure which Edwards gives from the life, it muſt be thicker*.

* *La Linotte-Brune* of Buffon, the *Fringilla Atra* of Gmelin, and the *Fringilla Obſcura* of Latham. "It is of a blackiſh duſky "colour, the breaſt and rump cinereous, the feathers lighter on its "crown." It inhabits Angola, and is four inches long.

M

IV. The

IV.

The DUSKY FINCH.

Our knowledge of this bird is drawn from Edwards. Almoſt all its feathers are blackiſh, edged with a lighter colour, which has a reddiſh caſt on the upper-part of the body: the general colour that reſults from this mixture is deep brown, though varied. It has a cinereous tint on the breaſt and rump; the bill is alſo cinereous, and the legs are brown.

I preſume that Briſſon ought not to confound this bird with *the Little Brown Sparrow* of Cateſby, whoſe plumage is of an uniform brown, without any moulting, and therefore quite different. But the difference of climate is ſtill greater; for Edwards's Duſky Linnet came probably from Brazil, perhaps even from Africa; whereas Cateſby's *Little Brown Sparrow* is found in Carolina and Virginia, where it breeds and continues the whole year. Cateſby tells us, that it lives upon inſects, that it is almoſt always alone, that it is not very common, that it viſits inhabited places, and that it is conſtantly hopping among the buſhes. We are not acquainted with the habits of the Duſky Linnet.

The MINISTER.

Le Miniſtre, Buff.
Tanagra Cyanea, Linn.
Emberiza Cyanea, Gmel. and Lath.
Tanagra Carolinenſis Cœrulea, Briſſ.
Blue Linnet, Edw.
The Indigo Bunting, Penn. and Lath.

THIS is the name given by bird-fanciers to a Carolina bird, which others call the *Biſhop*, but which muſt not be confounded with the Biſhop of Brazil, which is a Tanagre. I place it near the Linnets, becauſe in the time of moulting it is ſo much like theſe, as to be miſtaken for them, and the female at all ſeaſons reſembles them. The moult happens in the months of September and October; but the period varies as in Buntings, and in many other birds: the Miniſter is ſaid even to drop its feathers thrice a-year, in which circumſtance it alſo reſembles the Buntings, the Bengals, and Senegals, &c.

When clothed in its fineſt attire, it is ſky-blue upon a violet ground; the vane of the wings is of a deep blue, with deep brown in the male, and with a greeniſh tint in the female; which is ſufficient to diſtinguiſh it in the moulting ſea-

Its ſpecific character:—"It is azure, deeper coloured on the "crown, the quills of the wings and of the tail duſky, and edged "with cœrulean." It ſeems to be extremely like the Painted Bunting. *Emberiza Ciris* of Linnæus.

ſon

ſon from the male, whoſe plumage is in other reſpects pretty like that of the female.

The Miniſter is of the ſize of a Canary-finch, and, like it, lives upon millet, and the ſeeds of Canary-graſs, &c.

Cateſby figures this bird by the name of *the Blue Linnet* *, and tells us that it is found among the mountains in Carolina, at the diſtance of one hundred and fifty miles from the ſea; that its wings are nearly like the Linnet; that the feathers of its head are of a deeper blue, and thoſe of the under-part of the body of a lighter blue; that the quills of the tail are of the ſame brown with thoſe of the wings, with a light tinge of blue; and laſtly, that the bill is blackiſh and the legs brown, and that it weighs only two ounces and a half.

Its total length five inches; its bill five lines; the *tarſus* eight or nine lines; the middle toe ſix lines and a half; the tail two inches, and extends ten or twelve lines beyond the wings.

* The Spaniards call it *Azul Lexos*, or the far-fetched bird, as it comes to Mexico from the back parts of Carolina. It is ſmaller than the Goldfinch, and has the ſong of the Linnet. It appears in the ſtate of New York about the beginning of ſummer. It builds in the fork of a branch, with dry yellow graſs, and lines the neſt with the down of plants.

The BENGALS, and the SENEGALS.

Les Bengalis, et les Senegalis *, &c.

ALL travellers, and after them the naturalifts, have agreed, that thefe fmall birds change their colours in moulting. Some even add interefting particulars which we fhould wifh to afcertain; that the varieties of plumage are confined entirely to five principal colours, black, blue, green, yellow, and red; and that the Bengals never affume more than one at a time. Yet perfons who have had an opportunity of obferving thefe birds in France, and of watching their alterations for a courfe of years, affert that they have only one annual moult, and never change their colour †. This apparent contradiction may be explained by the difference of climates. That of Afia and Africa, the native regions of thefe birds, is more powerful than ours, and may have greater influence upon their plumage. But the Bengals are not the only birds which undergo the change; for, according to Merolla, the Sparrow in Africa becomes red in the rainy feafon,

* Some of them are termed the Senegal Sparrows.

† M. Mauduit, well known for his enlightened tafte in Natural Hiftory, and for his excellent collection of birds, obferved a Red Senegal that lived more than a year without changing his plumage. Château fays the fame of all the Bengals which paffed through his hands.

and

and many others are ſubject to ſimilar alterations. At any rate, an arrangement founded on the colours of the plumage muſt in the preſent caſe be totally uncertain; ſince, in their native climate, at leaſt, theſe pretended characters would only be momentary, depending on the ſeaſon when the birds was killed. On the other hand, their complexion, which fluctuates in Africa and Aſia, is invariable in Europe; and it becomes difficult to avoid compounding varieties with ſpecies. I ſhall follow therefore the received opinions, and allot a ſeparate article to each bird which appears obviouſly diſtinct, but without pretending to decide the number of real ſpecies, which can be only the work of time.

We ſhould be much miſtaken if we inferred from the names of theſe birds, that they are confined to Bengal and Senegal: they are ſpread through the greateſt part of Aſia and Africa, and even in many of the adjacent iſlands, ſuch as thoſe of Madagaſcar, Bourbon, France, and Java, &c. We may even expect ſoon to find them multiplied in America; for De Sonini lately ſet at liberty a great number of them in the Iſle of Cayenne, and afterwards ſaw them cheerful and lively, ſtrongly diſpoſed to naturalize in a foreign ſoil, and perpetuate their race *. We may hope that theſe new coloniſts, whoſe plumage is ſo variable, will alſo ſuffer the influ-

* A few years ago a Red Senegal was killed at Cayenne in a ſavanna; it had certainly been carried thither by ſome navigators.

ence

ence of an American climate, and other varieties will be produced, more fit however to decorate our cabinets than to enrich Natural Hiſtory.

The Bengals are familiar and deſtructive birds: in a word, they are real Sparrows. They viſit the houſes and even appear in the middle of the villages, and alight in numerous bodies in the fields ſowed with millet *; for they prefer this grain: they are alſo fond of bathing.

In Senegal, they are caught under a calebaſh, or large gourd, which is placed upon the ground, and raiſed a little by a ſhort prop, to which a long cord is faſtened; a few grains of millet ſerve for the bait. The perſon who watches their motions without being perceived, draws at the proper time, and ſecures whatever is under the calebaſh; Bengals, Senegals, and ſmall black Sparrows with white bellies, &c. † Theſe birds are tranſported with difficulty, and can hardly be reconciled to a different climate; but when once naturalized, they live ſix or ſeven years, that is longer than many ſpecies proper to the country. They have even bred in Holland; and the ſame ſucceſs would undoubtedly be had in

* Travellers inform us, that the negroes eat certain ſmall birds entire with their feathers, and theſe birds reſemble Linnets. I imagine that Senegals are of the number, for ſome Senegals in the time of moulting are like Linnets. Beſides, it is ſaid that the negroes eat the ſmall birds entire, only to retaliate for the damage done among their crops, theſe neſtling in the midſt of the ſowed fields.

† I owe the detail of this ſport to M. de Sonini.

in colder places; for these birds are very gentle and social, and often caress each other, and above all, the males and females sit near one another, and sing in concert. It is added, the song of the female is not much inferior to that of the male.

M

The BENGAL.

Le Bengali, Buff.
Fringilla Benghalus, (mas.) Linn.
Fringilla Angolensis, (fem.) Id.
The Blue-bellied Finch *, (fem.) Edw. and Lath.

As the instincts and habits are nearly the same in the whole of this family of birds, I shall content myself, in this and the following articles, to subjoin to the general account, the delineation of the peculiar features of each. In cases such as the present, where the principal object is to convey ideas of the richness and variety of the plumage, we ought to quit the pen for the pencil; at least, we must imitate the painter in describing not only the forms and lineaments,

* The specific character:—"Light cœrulean, head and back "gray, sides of the head purple." Brisson describes it:—"Gray "above, light cœrulean below; a purple spot below the eyes; "the rump and tail feathers light cœrulean." Bird-fanciers call it *Mariposa*.

but

but in reprefenting the fportive fluctuation of tints, their changing fucceffion and combination, and above all in expreffing action, motion, and life.

The Bengal has on each fide of its head a fort of purple crefcent which lies under the eyes, and marks the phyfiognomy of this little bird.

The throat is light blue, and the fame colour fpreads over all the lower parts of the body, as far as the end of the tail, and even over the upper coverts. All the upper-part of the body, including the wings, is of a pleafant gray.

In fome individuals, this fame gray, fomewhat lighter, is the colour of the belly and of the lower coverts of the tail.

In other fpecimens brought from Abyffinia, this gray had a tinge of red on the belly.

Laftly, in others there is no purple crefcent under the eyes; and this variety, known by the name of *Blue Cord* *, is more common than that firft defcribed. It is faid to be a female; but, as it is fo frequent, I fhould rather fuppofe that the appearance is owing to the difference of age or climate. Mr. Bruce, who has feen this bird in Abyffinia, pofitively affures us, that the two red fpots are not found in the female, and that all her colours are lefs brilliant. He adds, that

* *Cordon Bleu*, Knight of the order of the Holy Ghoft.

the male has an agreeable warble, but he never remarked that of the female: in both, the bill and the legs are reddiſh.

Edwards has figured and coloured a *Blue Cord (the Blue-bellied Finch)* which was brought from the coaſt of Angola, where the Portugueſe called it *Azulinha*. It differed from the preceding; the upper-part of the body being of a brown cinereous, ſlightly tinged with purple, the bill of a deep fleſh colour, and the legs brown. The plumage of the female was of a cinereous brown, with a ſlight tinge of blue on the lower-part of the body only. It would ſeem to be a variety from the climate, in which neither ſex has the red ſpot under the eyes; and this accounts for the frequency of the *Blue Cords*. It is a very lively bird. Edwards remarks that its bill is like that of the Goldfinch. He takes no notice of its ſong, not having an opportunity of hearing it.

The Bengal is of the ſize of the *Sizerin*; its total length is four inches nine lines; its bill four lines; its tail two inches, unequally tapering, and conſiſting of twelve quills; its extent ſix or ſeven inches.

M

The BROWN BENGAL*.

Brown is the predominant colour of this bird; but it is more intenſe under the belly, and mottled on the breaſt, with whitiſh in ſome individuals, and reddiſh in others. All the males have ſome of the upper coverts of the wings tipt with a white point, which produces a ſpeckling, confined however to that ſex; for the female is of an uniform brown without ſpots. In both the bill is reddiſh, and the legs of a light yellow.

The Brown Bengal is nearly the ſize of a wren; its total length is three inches and three-fourths; its bill is four lines; its alar extent about ſix inches and a half; and its tail rather more than an inch.

* Latham reckons this bird a variety of the Punctured Bengal, *Fringilla Amandava*, Linn. of the following article. Briſſon terms it *Bengalus Fuſcus*, or Duſky Bengal, and deſcribes it as "having "the throat and breaſt mixed with dirty whitiſh; the upper co-"verts of the wings dotted with white; the quills blackiſh."

M

The

N.o 89

FIG1 THE ANADUVADE FIG2 THE SENEGAL FINCH

The PUNCTURED BENGAL*.

Le Bengali Piqueté, Buff.
Fringilla Amandava, Linn.
Fringilla Rubra Minima, Klein.
Bengalus Punctulatus, Briff.
The Amaduvade Finch, Will. Alb. Edw. and Lath.

Of all the Bengals which I have feen, this is the moft fpeckled. The whole upper part of the body, the fuperior coverts of the tail and wings, and the quills of the wings next the back, were fprinkled with points; the wings were brown, and the lateral quills of the tail black, edged with white. Brown mixed with a dull red was fpread over all the upper part of the body, including the coverts of the tail, and even under the belly; a red not fo dufky extended over all the reft of the under-part of the body, and the fides of the head; the bill was alfo of a dull red, and the legs of a light yellow.

The female, according to Briffon, is never punctured; it differs alfo from the male, as its neck, breaft, and belly, are of a pale-yellow, and the throat white. According to other obfervers, who had many opportunities of

* Specific character:—" Dufky and tawnyifh, fpeckled with white; " the tail-quills black, with a white point at their tip." Briffon's defcription:—" Dufky above, mixed with dull red, below faintly " red; the upper coverts of the wings and of the tail, the breaft, " and the fides, fprinkled with white points; the feathers of the " wings black."

repeatedly ſeeing theſe birds alive, the female was entirely brown, and without ſpots. Is this a variety of plumage, or a difference of deſcription? for the latter is no ſmall ſource of confuſion in natural hiſtory. Willughby ſaw ſeveral of theſe birds which came from the Eaſt Indies, and, as we may expect, found ſeveral differences among the individuals; ſome had black wings; in others the breaſt was of that colour; in others the breaſt and belly were blackeſt; in others the legs whitiſh: in all the nails were very long, and more incurvated than thoſe of the lark. There is reaſon to believe that ſome of theſe birds were in moult; for I obſerved one in which the lower-belly was blackiſh, and all the reſt of the plumage indecided, as happens when the feathers are dropped, the colours peculiar to the ſpecies being impreſſed, but not well diſcriminated.

The ſpecimen deſcribed by Briſſon came from the iſland of Java. Thoſe obſerved by Charleton were brought from India; their warble was very pleaſant; ſeveral of them were kept together in the ſame cage, becauſe they diſliked the ſociety of other birds.

The Speckled Bengal is of a ſize intermediate to the two preceding; its total length is about four inches; its bill four or five lines; its extent leſs than ſix inches; its tail an inch and four lines, unequally tapered, and conſiſting of twelve quills.

The SENEGAL*.

Le Senegali, Buff.
Fringilla Senegala, Linn.
Senegalus Ruber, Briss.
The Senegal Finch, Lath.

THERE are two principal colours in the plumage of this bird; vinous red on the head, on the throat, and on all the under-part of the body as far as the legs, and on the rump; a greenish-brown on the lower belly and on the back, where it receives a slight tinge of red. The wings are brown, the tail blackish, the legs gray, the bill reddish, except the upper and lower ridge, and its edges, which are brown, and form a sort of red-coloured squares.

This bird is somewhat smaller than the Punctured Bengal, but longer shaped. Its total length is four inches and a few lines; its bill four lines; its alar extent six inches and a half; its tail eighteen lines, consisting of twelve quills.

* Specific character:—"Dusky ferruginous; rufous on the head, "and under; the bill red, streaked with black."—"It is greenish-"dusky above, mixed with wine colour, and below of a red wine "colour; the crown red wine; the lower belly greenish-dusky; the "quills of the wings black." BRISSON. Some have given it the name of *Ruby*, on account of its colour.

VARIETIES of the SENEGAL.

I. I have ſeen one of theſe birds which had been killed in Cayenne in a ſavanna, and the only one that has been ſeen in that country. It probably had been carried thither by ſome curious perſon, and had eſcaped from its cage. It differed in ſome reſpects from the preceding; the coverts of its wings were ſlightly edged with red; the bill was entirely of that colour; the legs only reddiſh: and what ſhews the cloſe analogy between the Bengals and Senegals, the breaſt and ſides were ſprinkled with ſome white points.

II. The Danbik of Mr. Bruce. This bird is very common in Abyſſinia, and partakes of the qualities of the two preceding. It is of the ſame ſize. The red colour, which is ſpread over all the anterior part, does not deſcend to the legs, as in the Senegal, but it extends over the coverts of the wings, where a few white points are perceived likewiſe on the ſides of the breaſt. The bill is purple, its upper and lower ridge bluiſh, and the legs cinereous. The male ſings agreeably. The female is of an almoſt uniform brown, and has very little purple.

The RADIATED SENEGAL*

Le Senegal Rayé, Buff.
Loxia Aſtrild, Linn.
Fringilla Undulata, Pall.
Senegalus Striatus, Briſſ.
Waxbill Groſbeak, Edw. and Lath.

It is radiated tranſverſely as far as the end of the tail with brown and gray, and the *ſtriæ* are the more delicate as they are nearer the head. The general complexion is much lighter on the lower part of the body; it is alſo ſhaded with roſe-colour, and there is a red oblong ſpot under the belly; the lower coverts of the tail are black without any rays, but ſome traces can be perceived on the wing-quills, which are brown; the bill is red, and there is a ſtripe, or rather a bar of that colour on the eyes.

I have been aſſured that the female is exactly like the male; but the differences which I have myſelf obſerved in many individuals, and thoſe which have been noticed by others, create ſome

* Linnæan ſpecific character:—" Gray, waved with duſky; the " bill, orbits, and breaſt, ſaffron-coloured."—" It is ſtreaked " tranſverſely with duſky and dirty gray, mixed with roſe-colour " in the lower part of the body, and with red on the belly; a red " ſtripe near the eyes; the quills of the wings ſtreaked tranſverſely " with duſky and dirty gray." Brisson.—It is called Waxbill on account of the colour of its bill. Some have confounded it with the *La-ki* of China, to which many marvellous properties are imputed; but that bird is as large as a blackbird, and bears no relation to the Senegals,

doubts

doubts of the perfect similarity of the sexes. I have seen several which came from the Cape, in some of which the upper-part of the body was more or less of a deep brown, and the under more or less reddish; in others the upper-part of the head had no rays. In that figured by Edwards, Pl. 179, the rays consisted of two browns; and the coverts below the tail were not black, which is also the case with what we have represented N° 157, fig. 2. Lastly, in the specimen delineated at the top of Pl. 354, the rays of the under-part of the body are spread upon a brown ground; and not only the lower coverts of the tail are black, as in that described by Brisson, but the lower belly is of the same colour.

The subject observed by Brisson came from Senegal. The two which Edwards examined were brought from the East Indies; and most of those which I have seen were brought from the Cape of Good Hope. Among so many differences of the plumage remarked between these, some must depend upon the distinction of sex.

The average length of these birds is about four inches and a half; the bill is three or four lines; the alar extent six inches, and the tail two inches, unequally tapered, and composed of twelve quills.

The SEREVAN.

Brown is the predominant colour of the head, the back, the wings, and the quills of the tail; the under-part of the body is light gray, ſometimes a light fulvous, but always tinged with reddiſh; the rump is red, and alſo the bill; the legs are red; ſometimes the baſe of the bill is edged with black, and the rump ſprinkled with white points, and ſo are the coverts of the wings. Such was the Serevan ſent from the Iſle of France by Sonnerat, under the name of *Bengal.*

That called *Serevan* by Commerſon had all the under-part of the body of a light fulvous; its legs were yellowiſh; and neither its bill nor its rump was red, and not a ſingle ſpeck could be ſeen on it. Probably it was young, or was a female.

Other birds cloſely related to this, and ſent by Commerſon, under the name of *Cape Bengals*, had a more diſtinct red tinge before the neck, and on the breaſt; in general their tail was longer in proportion.

They are all nearly of the ſize of the Bengals and Senegals.

The LITTLE SENEGAL SPARROW.

The bill and legs are red, and there is a ſtreak of the ſame colour on the eyes; the throat, and the ſides of the neck, are of a bluiſh white; all the reſt of the under-part of the body is white, mixed with roſe-colour of greater or leſs intenſity; the rump of the ſame; the reſt of the under-part of the body blue; the upper-part of the head is of a lighter blue; the wings, and the ſcapular feathers, brown; the tail blackiſh.

This Little Sparrow is nearly the ſize of the preceding.

The MAIA*.

Le Maia, Buff.
Fringilla Maia, Linn. Gmel. Briſſ. and Will.
The Cuba Finch, Lath.

Theſe are very deſtructive little birds. They aſſemble in numerous flocks to feed among the crops of rice; they conſume much, and waſte more;

* Specific character:—" Purpliſh, with a ſtripe on the breaſt of " a deeper colour."—Briſſon's deſcription:—MALE. " Cheſnut " purple above, blackiſh below; head and neck blackiſh; tranſverſe " ſtripe on the breaſt of a cheſnut purple; the wing-quills cheſ- " nut-purple above, and duſky verging on rufous below." FEMALE. " Fulvous above, of a dirty yellowiſh white below; the " throat, and a ſpot on either ſide the breaſt, of a cheſnut-purple; " the wing-quills fulvous."

It is four inches and three-fourths long.

N°. 90

THE CUBA FINCH

more; they prefer the countries where that grain is cultivated, and might claim with the *Paddas* the appellation of *Rice-birds*: however, I retain their proper name, by which, as Fernandez informs us, they are known in their native climate. The ſame author ſays, that their fleſh is good, and eaſy to digeſt.

In the male, the head, throat, and all the under-part of the body, are blackiſh; the under of a purple cheſnut, brighteſt on the rump; there is alſo on the breaſt a broad cincture of the ſame colour; the bill is gray, and the legs lead-coloured.

The female is fulvous above, and dirty white below; its throat is purple-cheſnut, and on each ſide of the breaſt is a ſpot of the ſame colour, correſponding to the cincture of the male; its bill is whitiſh, and its legs are gray.

Fernandez mentions as a *wonder*, that the ſtomach of the Maia is behind its neck; but if he had noticed the ſmall birds that are fed, he would have perceived that this wonder is very common; and that as the craw is filled, it is puſhed to the place where it meets with leaſt reſiſtance, often on the ſide of the neck, and ſometimes behind it: laſtly, he would have ſeen that the craw is not the ſtomach.—Nature is always admirable, but we ought to know how to admire her.

The MAIAN*.

Le Maian, Buff.
Loxia Maia, Linn. Gmel. and Briſſ.
The Malacca Groſbeak, Edw.
The White-headed Groſbeak, Lath.

China is not the only country from which this bird is brought; that engraved by Edwards came from Malacca, and in all probability it might be found in the intermediate countries. There is reaſon however to doubt whether it exiſts in America, and whether ſo ſmall a bird could traverſe the immenſe oceans which diſjoin the continents; at leaſt it differs ſo much from the Maias, the American birds which it the neareſt approaches, as to require a diſtinct name. In fact its properties are not the ſame; for though it be rather larger, it has its wings and tail ſomewhat ſhorter, and its bill as ſhort; beſides, its plumage is different, and much inferior in luſtre.

In the Maian, all the upper-part of the body is a reddiſh cheſnut; the breaſt, and all the under-part of the body, is of an almoſt uniform blackiſh, not quite ſo deep beneath the tail; the bill is of a lead-colour; a ſort of light gray cowl

* Its ſpecific character:—"Duſky, head white."—"It is above "of a duſky-cheſnut, below blackiſh; head and neck dirty white; "the breaſt faintly duſky; the wing-quills of a deep duſky-cheſ-"nut." Brisson.

covers

covers the head, and falls below the neck; the lower coverts of the wings are of the ſame light gray, and the legs are fleſh-coloured.

The Maian of Briſſon differs from this; its breaſt being of a light brown, ſome of the firſt feathers of the wings being edged with white, the bill and legs being gray, &c. Theſe differences are too ſtriking to be imputed to the variety of deſcription, eſpecially if we conſider the ſcrupulous accuracy of the deſcribers.

The CHAFFINCH.

Le Pinson, Buff.
Fringilla Cœlebs, Linn. and Gmel.
Fringilla, Gesner, Aldrov. and Briss. &c.
The Snowfleek, or Shoulfall *, Sib. Scot. Ill.

THIS bird has great power in its bill, with which it nips so bitterly as to draw blood. Hence, according to the several authors, the French name of *Pinson* is derived (from *pincer*, to pinch). But as the disposition to snap is not peculiar to the Chaffinch, but common to other birds, to many different sorts of quadrupeds, millepedes, &c. I should rather adopt the opinion of Frisch, who supposes this word *Pinson* to come from *Pincio*, latinized for the German *Pinck*, which seems imitative of the note of the bird.

The Chaffinches do not all migrate in the autumn; a considerable number remains with us during the winter. These resort to our dwellings, and even visit our court-yards to obtain an easier subsistence; they are little parasites, who seek to live at our expence, without contri-

* Aristotle calls the Chaffinch Σπιζα: the Italian names are, *Franguello*, *Francueglio*, and *Frenguello*: in German, *Finck*, *Roth Finck*, *Winche*: in Polish, *Slowick*: in Swedish, *Finke*, *Bofincke*: in Bohemian, *Penkewa*.

buting

N.º 91

THE CHAFFINCH

buting to our pleaſures: for in that ſeaſon they never ſing, except in fine days, which ſeldom then occur. During the reſt of the time they lodge concealed in cloſe hedges, in oaks that have not ſhed all their leaves, in evergreens, and even in holes of rocks, where they are ſometimes found dead when the weather is exceſſively ſevere. Thoſe which migrate into other climates aſſemble often in prodigious flocks; but whither do they retire? Friſch ſuppoſes that their retreat lies in the northern regions: his reaſons are; 1. That on their return, they bring with them white Chaffinches, which are hardly to be met with, except in thoſe climates; 2. That they never lead young ones in their train, which would be the caſe, if, during their abſence, they lived in a warm country, where they might be diſpoſed to breed; 3. That they can bear cold, except ſnow, which covering the fields, deprives them in part of their ſubſiſtence *.

Theſe arguments imply therefore, that there is a country in the north where the earth is not robed in the winter; and ſuch is ſaid to be the

* Friſch.—Aldrovandus ſays, that in Italy, when there is much ſnow and the froſt very intenſe, the Chaffinches cannot fly, and are caught by the hand; but this weakneſs may ariſe from inanition, and that again from the abundance of ſnow. Olina alleges, that in the ſame country the Chaffinches retire to the hilly tracts during the ſummer. Hebert has ſeen many of them in that ſeaſon among the higheſt mountains of Bugey, where they were as common as in the plains, and where they certainly do not remain through the winter.

desarts of Tartary, where the snow falls indeed, but is immediately swept away by the violence of the winds, and extensive tracts laid bare.

A very remarkable circumstance in the migration of the Chaffinches, is what Gesner mentions of those of Switzerland, and Linnæus of those of Sweden: that only the females remove to other climates, and the males reside in the country during the winter *. But have not these able naturalists been misled by the persons who informed them, and those deceived by some periodical change in the plumage of the females, occasioned by cold, or other accidents? This explanation seems more suited to Nature's general plan, and more conformable to analogy, than to suppose that, at a certain fixed term, the females separate from mere caprice, and travel into a distant climate, when their native soil can afford them subsistence.

The migrations vary in different countries. Aldrovandus assures us, that the Chaffinches seldom breed near Bologna, but almost all depart

* "They quit Switzerland in the winter, particularly the females; for several males are often seen, with not one female." GESNER. Linnæus positively asserts, that the female Chaffinches leave Sweden in flocks about the month of September, that they direct their course to Holland, and return in the spring to join their males which have wintered in Sweden.

This singular remark is corroborated by Mr. White, who found that the vast flocks of Chaffinches which appear in his neighbourhood about Christmas, are almost entirely hens. Yet, as he seems not to have dissected any, he might have been deceived by the change of plumage, which is extremely probable.

about the end of winter, and return the ſucceeding autumn. On the other hand, I find from Willughby, that they continue the whole year in England, and that few birds are ſo common.

They are ſpread through the whole of Europe, from the ſhores of the Baltic and Sweden, where they are frequent, and neſtle, to the Straits of Gibraltar, and even the coaſts of Africa *.

The Chaffinch is a lively bird, and perpetually in motion; and this circumſtance, joined to the ſprightlineſs of its ſong, has given riſe to the proverbial ſaying, *gay as a Chaffinch.* It begins to chant very early in the ſpring, and ſeveral days before the Nightingale, and gives over about the ſummer ſolſtice. Its ſong has merited an analyſis: and it is diſtinguiſhed into the prelude, the quavering, and the cloſe †; names have been appropriated to the different parts; and the greateſt connoiſſeurs in theſe

* "Being ſtationed on the coaſts of Africa, we were very "often viſited on board by Chaffinches. We cruiſed between the "thirtieth to the thirty-fifth degree of latitude:—I have even "heard it affirmed, that they are found at the Cape of Good "Hope." Note of *Viſcount Querhoent.*

† The prelude is, according to Friſch, compoſed of three ſimilar notes or ſtrokes; the quaver, of ſeven different notes deſcending; and the cloſe, of two notes. Lottinger has alſo made ſome obſervations on this ſubject. "In anger, the air of the "Chaffinch is ſimple and ſhrill; in fear, plaintive, ſhort, and "often repeated; in joy, it is lively, and ends with a ſort of "burden."

 little

little matters agree, that the concluding part is the moſt agreeable *. Some find its muſic too ſtrong, or too grating †; but this muſt be imputed to the exceſſive delicacy of our organs, or rather it is becauſe the ſound is too near, and increaſed by the confuſed echo of our apartments: Nature has deſtined the Chaffinches to be ſongſters of the woods; let us repair then to the grove, to taſte and enjoy the beauties of their muſic.

If a young Chaffinch taken from the neſt be educated under a Canary, a Nightingale, &c. it will have the ſong of its inſtructors: more than one inſtance ‡ has been known of this; but they have never been brought to whiſtle our tunes:—they never depart ſo wide from nature.

The Chaffinches, beſide their ordinary warble, have a certain tremulous expreſſion of love, which they can utter in the ſpring, and alſo another cry which is unpleaſant, and ſaid to portend rain §. It has been remarked too, that they never ſing better or longer than when, from ſome accident, they have loſt their ſight ‖; and

* In German, this is called *Reiterzu*; in French, *Boute-ſelle*.

† *Mordant*, biting.

‡ This facility in learning the ſongs of other birds accounts for the diverſity obſerved in the warble of the Chaffinches. In the Netherlands, five or ſix kinds of Chaffinches are diſtinguiſhed by the various length of their airs.

§ In the German language a word is appropriated to denote this: it is *Schircken*.

‖ They are liable to this accident, eſpecially if kept between two windows which face the ſouth.

no

no ſooner was this obſervation made, than the art was diſcovered of rendering them blind. The lower eyelid is connected to the upper by a ſort of artificial cicatrix made by touching ſlightly and repeatedly the edges with a wire heated red-hot in the fire, and taking care not to hurt the ball of the eye. They muſt be prepared for this ſingular operation by confining them for ten or twelve days to the cage, and then keeping them ſhut up with the cage in a cheſt night and day, to accuſtom them to feed in the dark*. Theſe blind Chaffinches are indefatigable ſingers, and they are preferred as calls to decoy wild Chaffinches into the ſnares: theſe are alſo caught with bird-lime and with different kinds of nets, and among others thoſe for larks, but the meſhes muſt be ſmaller in proportion to the ſize of the bird.

The time for the ſport is, when the Chaffinches fly in numerous flocks, either in autumn before their departure, or in ſpring on their return. We muſt, as much as poſſible, chooſe calm weather, for they keep lower and hear better the call. They do not eaſily bend to captivity; they ſcarcely will eat any thing for the firſt two or three days; they ſtrike their bill

* Geſner aſſerts, that if the Chaffinches be kept thus ſhut up through the whole ſummer, and not let out of their priſon till the beginning of autumn, they ſing during the latter ſeaſon, which would otherwiſe not happen. Darkneſs rendered them dumb, but return of light is to them a ſecond ſpring.

 continually

continually againſt the ſticks of the cage, and often languiſh to death *.

Theſe birds conſtruct their neſt very round and compact, and place it in the cloſeſt trees or buſhes; ſometimes they build it even in our gardens upon the fruit-trees, and conceal it ſo artfully that we can hardly perceive it, though quite nigh. It is compoſed of white moſs and ſmall roots on the outſide, and lined with wool, hairs, ſpiders-webs, and feathers. The female lays five or ſix eggs, which are reddiſh gray, ſprinkled with blackiſh ſpots, more frequent near the large end. The male never deſerts his mate in the time of hatching; he ſits at night always at hand; and if during the day he remove to a ſhort diſtance, it is only to procure food. Jealouſy has perhaps ſome ſhare in this exceſſive aſſiduity; for theſe birds are of an amorous complexion: when two males meet in an orchard in the ſpring, they fight obſtinately, till one of them is vanquiſhed and expelled: and the combat is ſtill more fierce if they be lodged in the ſame volery with only a ſingle female.

The parents feed their brood with caterpillars and inſects: they alſo eat theſe themſelves, but their ordinary ſubſiſtence is ſmall ſeeds, thoſe of the white thorn, of poppy, of burdock, of the roſe-tree, and eſpecially beech-maſt, rape and

* Thoſe caught with lime-twigs often die the inſtant they are taken.

hemp

hemp ſeed. They feed alſo upon wheat and even oats, and are expert in ſhelling the grain to obtain the mealy ſubſtance. Though rather obſtinate, they can in time be inſtructed like Goldfinches to perform ſeveral little feats; they learn to employ their wings and feet to draw up the cup when they want to eat or drink.

The Chaffinch ſits oftener ſquatted than perched; it never walks hopping, but trips lightly along the ground, and is conſtantly buſy in picking up ſomething: its flight is unequal; but when its neſt is attacked, it hovers above ſcreaming.

This bird is ſomewhat ſmaller than our Sparrow, and is too well known to require a minute deſcription. The ſides of the head, the fore-part of the neck, the breaſt, and the loins are of a wine colour: the upper-part of the head and of the body cheſnut; the rump olive, and a white ſpot on the wing. In the female the bill is more ſlender, and the colours leſs bright than in the male; but in both ſexes the plumage is very ſubject to vary. I have ſeen a hen Chaffinch alive, caught on her eggs the 7th of May, which differed from that deſcribed by Briſſon: the upper-part of its head and back was of a brown olive, a ſort of gray collar ſurrounded the neck behind, the belly and the lower coverts of the tail were white, &c. And of the males, ſome have the upper-part of the head and neck cinereous, and others of a brown chesnut; in some the quills of the tail nearest the two middle ones are edged with white, and in others they are entirely black. Does age occasion these slight differences?

A young Chaffinch was taken from its mother, when its tail-quills were six lines in length, and the under-part of its body was like that of its mother; and the upper-part of a brown cinereous; the rump olive; the wings were already marked with white rays: but the edges of the superior mandible were not yet scalloped near the point as in the adult males. This circumstance would lead me to suppose that the scalloping which occurs in many species is not the primary organization, but is afterwards produced by the continual pressure of the end of the lower mandible, which is rather shorter, against the sides of the upper.

All the Chaffinches have the tail forked, composed of twelve quills; the ground colour of their plumage is dull cinereous, and the flesh is not good to eat. The period of their life is seven or eight years.

Total length six and one-third inches; the bill six lines; the alar extent near ten inches; the tail two and two-thirds, and extends about sixteen lines beyond the wings. [A]

[A] Specific character of the Chaffinch, *Fringilla Cœlebs*, LINN. —" Its joints are black; its wing-quills white on both sides, the " three

" three first without spots, two of the tail-quills obliquely white." Thus described by Brisson:—" Above it is dusky-chesnut; below " white tawny; its rump green olive, (the lower-part of the neck " and the breast wine coloured in the male,) with a white spot on " the wings; the lateral quills of the tail are black, the outermost " distinguished by an oblique white stripe, the next terminated " obliquely with white on the inside."

M

VARIETIES of the CHAFFINCH*.

Before the frequent variations which may be perceived in Chaffinches bred in the same country, others are observed in different climates which are more permanent, and which authors have judged worthy of description. The three first have been found in Sweden, and the remaining two in Silesia.

I. The CHAFFINCH † with black wings and tail. The wings are indeed entirely black, but the outer quills of the tail, and the one next to it, are edged with white on the outside from the middle. This bird lodges among trees, says Linnæus.

* This Finch is termed *Fringilla Sylvia,* in the Fauna Suecica.
† *Fringilla Flavirostris Fusca,* Syst. Nat. Ed. x.

II. The

II. The BROWN CHAFFINCH *. It is distinguished by its brown colour and its yellowish bill, but the brown is not uniform, it is lighter on the anterior part, and has a shade of the cinereous and blackish of the posterior part. This variety has black wings like the preceding; the legs are of the same colour, and the tail forked. The Swedes call it *Riska*, according to Linnæus.

III. The CRESTED BROWN CHAFFINCH. It is fire-coloured, and this character distinguishes it from the preceding variety. Linnæus said in 1746, that it was found on the northern part of Sweden, but twelve years afterwards he recognized it to be the Black Linnet of Klein, and asserted that it inhabited every part of Europe.

IV. The WHITE CHAFFINCH †. It is very rare according to Schwenckfeld, and differs only in regard to colour from the Common Chaffinch. Gesner affirms, that a Chaffinch was seen whose plumage was entirely white.

* *Fringilla Flammea Fusca,* Syst. Nat. Ed. x.
† *Fringilla Candida,* Schwenckfeld.

V. The

V. The COLLARED CHAFFINCH †. The crown of its head is white, and it has a collar of the ſame colour ;—this bird was caught in the woods near Kotzna.

† *Fringilla Torquata*, Schwenckfeld.

The BRAMBLING.

Le Pinſon D'Ardenne, Buff.
Fringilla-Montifringilla, Linn. and Gmel.
Montifringilla, Geſner, Aldrov. Briſſ. &c.
Fringilla Montana, Roman. Orn.
The Bramble, or *Brambling,* Will.
The Mountain Finch, Ray.

PERHAPS this bird, which in general is ſuppoſed to be the Mountain Finch, or *Oroſpiza* of Ariſtotle, is in fact his *Spiza*, or principal Finch; or our Common Finch or Chaffinch is his Mountain Finch. The following are the reaſons which incline me to this opinion. The ancients never made complete deſcriptions, but ſeized a prominent feature of an animal, whether in its exterior appearance or in its habits, and marked it by an epithet. The *Oroſpiza*, ſays Ariſtotle *, is like the *Spiza*; it is ſomewhat ſmaller; its neck is blue; and laſtly, it inhabits the mountains; but all theſe are properties of the Chaffinch, and ſome of them belong to it excluſively.

1. It is much like the Mountain Finch or Brambling, as will appear from the compariſon; and all ſyſtematic writers have claſſed them together.

* It is ſuppoſed to be Ariſtotle's Ορεσπιζα, or Mountain Finch, whoſe female was termed χρυσομιτρις, or Golden Mitred. In German, *Rowert, Schnec-Finck, Winter-Finck:* in Swiſs, *Wald-Finck, Thann-Finck:* in Swediſh, *Norrquint.*

2. The

2. The Chaffinch is rather ſmaller than the Brambling, according to naturaliſts, and which agrees with my own obſervations.

3. In the Chaffinch, the upper-parts of the head and of the neck are of a bluiſh cinereous; whereas in the Brambling theſe are varied with gloſſy black and yellowiſh gray.

4. We have already remarked, on the authority of Olina, that in Italy the Chaffinch retires in ſummer to breed among the mountains; and, as the climate of Greece is little different from that of Italy, we may infer from analogy, it will there alſo have the ſame habits *.

5. Laſtly, the *Spiza* of Ariſtotle appears to reſort, according to that philoſopher, to the warm regions during ſummer, and to prefer the cold climates in winter †. But this agrees better with the Brambling than with the Chaffinches, ſince of theſe a great proportion never migrate, while the former not only are birds of paſſage, but

* Friſch aſſerts that the Bramblings come from the mountains in autumn, and when they return they direct their courſe to the north. The Marquis de Piolenc, who has given me ſeveral notes on theſe birds, aſſures me that they leave the mountains of Savoy and Dauphiny in October, and do not return till February. Theſe periods correſpond well with the time when they are ſeen to paſs and repaſs in Burgundy.—Perhaps both theſe ſpecies reſemble each other in preferring mountains.

† Aldrovandus poſitively aſſerts, that this takes place in the neighbourhood of Bologna: Lottinger informs me, that ſome appear in Lorraine from the end of Auguſt, but that large flocks arrive towards the end of October, and even later.

generally

generally arrive in the depth of winter * in the different countries which they viſit. This is evinced by experience, and is confirmed by the appellations of Winter-Finch, and Snow-Finch, which they have received in various places.

From all theſe conſiderations, it ſeems probable that the Brambling is the *Spiza* of Ariſtotle, and the Chaffinch his *Oroſpiza*.

The Bramblings do not breed in our climates; they arrive in different years in immenſe flocks. The time of their paſſage is the autumn and winter: often they retire in eight or ten days, and ſometimes they remain till the ſpring. During their ſtay, they conſort with the Chaffinches, and, like theſe, ſeek concealment in the thick foliage. Vaſt bodies of them appeared in Burgundy in the winter of 1774, and others in ſtill more numerous flocks were ſeen in the country of Wirtemberg about the end of December 1775, which every evening repoſed in a valley adjoining to the banks of the Rhine †, and commenced

* *Hiſt. Anim.* lib. viii. 3.

† Lottinger aſſerts perhaps too generally, that in the day-time they ſpread through the foreſts of the plain, and in the evening retire to the mountains. This conduct is not invariable, but ſeems to be affected by ſituation and circumſtances.

A flock of more than three hundred were ſeen this year in our neighbourhood; it halted three or four days in the ſame place, which is mountainous. They always alighted on the ſame cheſnut-tree, and when fired at, roſe all at once, and conſtantly directed their courſe to the north and north-eaſt.

Note of the Marquis PIOLENC.

their

their flight with the earlieſt dawn: the ground was covered with their excrements. The ſame occurrence was obſerved in the year 1735 and in 1757 *. Never perhaps were ſo many of theſe birds ſeen in Lorraine, as in the winter of 1765: more than ſix hundred dozen, ſays Lottinger, were killed every night in the pine-foreſts, which are four or five leagues from Sarbourg. The people were not at the pains to ſhoot them, they knocked them down with ſwitches; and though this maſſacre laſted the whole winter, the body was ſcarce perceptibly thinned. Willughby tells us, that many are ſeen in the neighbourhood of Venice, no doubt in the time of paſſage; but no where do they appear ſo regularly as in the foreſts of Weiſſemburg, which are plentifully ſtocked with beeches, and conſequently afford abundance of maſt, of which they are ſo fond, that they eat it day and night; they live alſo on all ſorts of ſmall ſeeds. I ſuppoſe that theſe birds remain in their native climate as long as they can procure the proper food, and quit it only when ſcarcity obliges them to ſhift their quarters; at leaſt, it is certain that the plenty of their favourite ſeeds is not ſufficient to draw them to a country, and even to one with which they are acquainted: for in 1774, when there was abundance of beech maſt in Lorraine, the Bramblings did not appear, but took a different

* *Gazette d'Agriculture*, Ann. 1776.

route:

route: however, in the following year, ſeveral flocks were ſeen, though there was a ſcarcity of maſt *. When they arrive among us, they are not ſhy, but allow a perſon to go very near them. They fly cloſe together, and alight and riſe in the ſame compact body; and for this reaſon twelve or fifteen of them may be killed at one ſhot.

When they feed in the fields, they are obſerved to perform the ſame manœuvres as the pigeons; a few always precede, and are followed by the reſt of the flock.

Theſe birds, we ſee, are known and ſpread through all parts of Europe; but they are not confined to our quarter of the globe. Edwards obſerved ſome that were brought from Hudſon's-bay, under the name of *Snow-birds;* and people who traded to that country aſſured him that they were the firſt which appeared every year on the return of ſpring, before even the ſnows were melted.

The fleſh of the Bramblings, though ſomewhat bitter, is good to eat, and undoubtedly better than that of the Chaffinch. Their plumage is alſo more varied, more beautiful, and more gloſſy; but their ſong is far from being ſo pleaſant, and it has been compared to the ſcreech of the owl † and the mewing of the cat ‡. They have two cries; the one a ſort

* I owe theſe facts to Mr. Lottinger. † Belon. ‡ Olina.

of

of chirping, and the other which they utter when they ſit on the ground reſembles that of the Stone-chat, but is neither ſo ſtrong nor ſo clear. Though by nature endowed with ſo few talents, theſe birds are ſuſceptible of inſtruction; and when kept near another whoſe warble is more pleaſant, their ſong gradually mellows, and comes to reſemble that which they hear*. But to have a juſt idea of their muſic, we muſt liſten to them in the time of hatching; it is then, when chanting the hymn of love, that birds diſplay their true warble.

A fowler, who had travelled, aſſured me that theſe birds are bred in Luxemburg; that they make their neſts in the moſt branchy firs at a conſiderable height; that they begin about the end of April; that they employ the long moſs of firs for the outſide, and hair, wool, and feathers for the lining; that the female lays four or five yellowiſh ſpotted eggs; and that they begin to flutter from branch to branch about the end of May.

The Brambling is, according to Belon, a courageous bird, and defends itſelf with its bill to the laſt gaſp. All agree that it is of a more eaſy temper than the Chaffinch, and more readily enſnared. Many of them are killed in certain fowling-matches which are frequent in the country of Weiſſemburg, and which deſerve to be re-

* Olina.

lated. The fowlers aſſemble at the little town of Bergzabern; on the evening of the day appointed, they diſpatch ſcouts to remark the trees on which the Bramblings commonly paſs the night, and which are generally the pitch-pines, and other ever-greens; the ſcouts, after their return, ſerve as guides for the company, which ſet out in the evening with torches and ſhooting-trunks *. The birds are dazzled with the glare, and killed by pellets of dry earth diſcharged from the trunks. They ſhoot very near, leſt they ſhould miſs; for if a bird chanced to be wounded, its cries would ſcare away the flock.

The principal food of theſe birds, when kept in a cage, is panic, hemp-ſeed, and beech-maſt. Olina ſays that they live four or five years.

Their plumage varies: in ſome males the throat is black; in others, the head is entirely white, and the colours in general lighter †. Friſch remarks, that the young males are not ſo black at their arrival, and that the inferior coverts of their wings are not ſo vivid a yellow as at their departure. Perhaps a more advanced age occaſions ſtill other differences between the ſexes, and may account for the diverſity of deſcriptions.

The Brambling which I obſerved weighed an ounce; its face was black; the upper-part of its

* *Sarbacanes.* † Aldrovandus.

head,

head, neck, and back, varied with yellowish-gray, and glossy-black; the throat, the fore-part of the neck, the breast, and the rump, rust-coloured; the small coverts of the base of the wing, yellow-orange; the others formed two transverse rays of a yellowish-white, separated by a broader black bar; all the quills of the wing, except the three first, had on their outer edge, where the great coverts terminate, a white spot, about five lines long; the succession of these spots formed a third white ray, which was parallel to the two others when the wing was expanded, but when the wing was closed it appeared only like an oblong spot almost parallel to the side of the quills; lastly, these quills were of an exceedingly fine black, edged with white. The small inferior coverts of the wings next the body were distinguished by their beautiful yellow colour. The quills of the tail were black, edged with white, or whitish; the tail forked; the flanks streaked with black; the legs of a brown-olive; the nails slightly incurvated, the hind one the strongest of all; the edges of the upper mandible scalloped near the point, the edges of the lower one fitted into the upper; and the tongue parted at the tip into several delicate filaments.

The intestinal tube was fourteen inches long; the gizzard was muscular, coated with a cartilaginous membrane slightly adhering, and preceded by a dilatation of the *œsophagus*, and also by a craw of five or ſix lines diameter: the whole was filled with ſmall ſeeds without a ſingle pebble. I did not ſee a *cæcum*, or gall-bladder.

The female has not the orange ſpot at the baſe of the bill, nor the fine yellow colour of the lower coverts; the throat is of a lighter rufous; and it has a cinereous caſt on the crown of the head, and behind the neck.

Total length ſix inches and one-fourth; bill ſix lines and a half; alar extent ten inches; tail two inches and one-third, and reaches about fifteen lines beyond the wings. [A]

[A] Specific character of the Brambling, *Fringilla Montifringilla*, LINN.—" The baſe of its wings very yellow below." Briſſon thus deſcribes it:—*Male.* " Above black, the margins of " the quills tawny, below white; the rump bright white; the " lower part of the neck and the breaſt dilute-rufous; the lateral " tail-quills blackiſh, their outer margins yellowiſh white, the outer-" moſt has its firſt half white exteriorly."—*Female:* " Above " duſky, the margins of its quills gray-tawny, below white; rump " bright white; lower part of the neck and the breaſt gray-tawny; " the lateral tail-quills duſky, their outer margins yellowiſh white, " the outermoſt has its firſt half white exteriorly."

The Bramblings ſometimes viſit Britain in winter.

M

The

The LAPLAND FINCH.

Le Grand-Montain, Buff.
Fringilla Lapponica, Linn.
Fringilla Montana, Briss. and Klein.
Fringilla Calcarata, Pall.
Montifringillæ Congener, Aldr.
Greater Brambling, Alb.
The Lapland Finch, Penn. and Lath.

This bird is the largest of the European Finches. Klein says that it is equal in bulk to the lark. It is found in Lapland, near Torneo. Its head is blackish, varied with a rusty-white colour, and marked on each side with a white ray, which rises from the eye, and descends along the neck; the neck, throat, and breast, are of a light-rufous colour; the belly, and the hind part, white; the upper-part of the body rusty, variegated with brown; the wings black, edged with pale-yellow and greenish, and crossed with a white ray; the tail forked, composed of twelve quills that are almost black, and edged with yellowish; the bill horn-coloured, and deeper near the point; the legs black.

Total length six inches and a half; bill seven lines, and the legs and mid-toe the same; alar extent eleven inches and a half; tail two inches and a half, and stretches ten lines beyond the wings. [A]

[A] Specific character of the *Fringilla Lapponica*, Linn.—"Its "head is black, its body gray and black, its eye-brows white, its "outermost tail-quills marked with a wedge-shaped white spot."

It inhabits Greenland in the ſummer, lays in June, and ſoon retires. It is found alſo in Lapland, and in the northern parts of Siberia. It appears in November at Hudſon's-bay, where it paſſes the winter among the juniper buſhes. It ſings nearly like the Linnet, but has a loftier and better ſupported flight. It trips on the ground like a Lark, picking up ſeeds.

The SNOW-FINCH.

Le Pinſon de Neige, ou *la Niverolle*, Buff.
Fringilla Nivalis, Linn. Gmel. and Briſſ.

This appellation is probably founded on the white colour of the throat, breaſt, and all the under-part of this bird; and alſo on the circumſtances of its inhabiting the cold countries, and ſcarcely appearing in temperate climates, except in winter when the ground is covered with ſnow. Its wings and tail are black and white; the head, and upper-part of the neck, cinereous, in which it reſembles the Chaffinch; the upper part of the body of a gray-brown, varied with lighter colour; the ſuperior coverts of the tail entirely black, and alſo the bill and legs.

Total length ſeven inches; the bill ſeven lines; the legs nine lines and a half; alar extent twelve inches; the tail two inches and ſeven lines, and ſtretches eight or nine lines beyond the wings. [A]

[A] Specific character of the *Fringilla Nivalis*:—"It is black, "below ſnowy, the ſecondary quills of the wing and the coverts "white." It is ſeven inches long.

The BROWN GROSBEAK.

*Le Brunor**, Buff.
Loxia Fuſca, Linn. and Gmel.

This is the ſmalleſt of all the Finches. Its throat, breaſt, and all the upper-part of the body of an orange reddiſh; the head, and all the under-part of the body, is of a deep brown; but the feathers are edged with a lighter ſhade, which produces a mixed colour; laſtly, the bill is white, and the legs brown.

Edwards, to whom we are indebted for our knowledge of this bird, could not diſcover from what country it came. Linnæus ſays that it is found in India.

Total length, three inches and one-fourth; bill, three lines and a half; legs, four lines and a half; tail, one inch, and extends ſix lines beyond the wings. [A]

* i. e. *Brun-noir,* or brown-black.

[A] Specific character of the *Loxia Fuſca*:—"It is duſky; "below whitiſh; the wing-quills from the third to the ninth are "entirely white." It inhabits Africa and Bengal. It is nearly of the bulk of a Canary.

The COWPEN FINCH.

Le Brunet, Buff.
Fringilla Pecoris, Gmel.
Fringilla Virginiana, Briss.

The prevailing colour of this bird is brown; but it is lighter under the body. Catesby tells us that it is an inhabitant of Virginia, and that it associates with the red wing orioles and the purple grakles: he adds, that it loves to haunt the cow-pens, and hence its name; and that it is never seen in summer.

Total length, six inches and three-fourths; the bill seven lines; the tail two inches and a half, and extends fifteen lines beyond the wings; the legs eleven lines, the mid-toe the same. [A]

[A] Specific character of the *Fringilla-Pecoris*:—"It is dusky, "below more dilute, the tail somewhat forked." It is larger than an English Bullfinch.

The BONANA FINCH.

Le Bonana, Buff.
Fringilla Jamaica, Linn. Gmel. Briss.
Passer Cæruleo-Fuscus, Ray, Sloan, Klein.
Emberiza Remigibus Rectricibusque Nigris, Amæn. Ac.
Gray Grosbeak, Brown's Jam.

This bird delights to perch on the banana, or bonana, which has given occasion to its name. The feathers of the upper-part of the body are ſilky, and dull blue; the belly variegated with yellow; the wings and tail of a dull blue, bordering on green; the legs black; the head large in proportion to the body; the bill ſhort, thick, and round.

This bird inhabits Jamaica.

Total length four inches and a half; the bill four lines; the alar extent eight inches and ſome lines; the tail about ſixteen lines, and ſtretches five or ſix lines beyond the wings. [A]

[A] Specific character of the *Fringilla Jamaica*:—"It is gray, "its breaſt green-cœrulean, the quills of its tail and wings black." It is of the bulk of the Siſkin, and five inches long.

M

The ORANGE FINCH.

*Le Pinſon a Tête Noire & Blanche**, Buff.
Fringilla Zena, Linn.
Fringilla Bahamenſis, Briſſ. and Klein.
The Bahama Finch, Cateſby.

The head, back, and ſcapular feathers, are black; but on each ſide of the head are two white rays, one of which paſſes above, and the other below the eye. The neck is black before and dull red behind, which is ſpread over the rump, and the ſuperior coverts of the tail; the throat is yellow; the breaſt, orange; the belly is white as far as the lower coverts of the tail,

* *i. e.* Black and White-headed Finch.

and

and including them; the tail is brown, and the wings are of the ſame colour, but have a white tranſverſe ray.

This bird is very common in Bahama, and in many other tropical parts of America. It is nearly of the ſize of the Common Chaffinch; it weighs ſix *gros*.

Total length ſix inches and one-fourth; the bill ſeven lines; the tail two inches and one-third, and extends about fifteen lines beyond the wings. [A]

[A] Specific character of the *Fringilla Zena*: – "It is black, "below white; a line above and below the eye bright white, the "breaſt fulvous."

M

The TOWHE BUNTING.

Le Pinſon Noir aux Yeux Rouges *, Buff.
Emberiza Erythrophthalma, Gmel.
Fringilla Erythrophthalma, Linn.
Fringilla Carolinenſis, Briſſ.

Black predominates on the upper-part of this bird (on the top of the breaſt, according to Cateſby) and on the quills of the wings, and the tail; the latter, however, are edged with white; the middle of the belly is white; the reſt of the under-part of the body dull red; the back black; the eyes red; and the legs brown.

* *i. e.* "Black Finch with red eyes:" The Linnæan appellation alſo of *Erythrophthalma* (ερυθρωφθαλμος) expreſsly the ſame.

The

The female is entirely brown, with a red tinge on the breaſt.

This bird is found in Carolina; it goes in pairs, and lodges in the thickeſt woods; it is of the bulk of a Creſted Lark.

Total length eight inches; the bill eight lines; the legs ſixteen lines; the tail three inches, and extends about twenty-ſeven lines beyond the wings, from which circumſtance we may infer that it cannot fly to a great diſtance. [A]

[A] Specific character of the *Emberiza Erythrophthalma*:—"It "is black ſhining with red, the lower belly tawny, with a white "ſpot on the wings."

M

The BLACK and YELLOW FINCH.

Le Pinſon Noir & Jaune, Buff.
Fringilla Capitis Bonæ Spei, Briſſ.

The general colour of this bird is velvet-black, which ſets off the beautiful yellow that prevails on the baſe of the wing, the rump, and the ſuperior coverts of the tail, and which borders the large quills of the wings. The ſmall quills, and the great coverts, are edged with gray; the bill and legs are alſo gray.

This bird was ſent from the Cape of Good Hope, and is of the ſize of an ordinary Chaffinch.

Total

Total length above ſix inches; the bill eight lines; the legs twelve lines; the mid-toe ten lines, the hind-toe nearly as long; alar extent ten inches and one-fourth; the tail two inches and two lines, and ſtretches twelve lines beyond the wings.

The LONG-BILLED FINCH.

Le Pinſon a Long Bec, Buff.
Fringilla Longiroſtris, Gmel.
Fringilla Senegalenſis, Briſſ.

The head and throat are black; the upper-part of the body varied with brown and yellow, the under-part with yellow-orange; it has a cheſnut collar; the quills of the tail are olive on the outſide, the great quills of the tail are of the ſame colour, tipt with brown; the middle ones brown, edged with yellowiſh; the bill and legs gray brown. It was ſent from Senegal. Its bulk nearly that of the Common Chaffinch.

Total length ſix inches and one-fourth; the bill nine lines; the legs eleven lines; the mid-toe ten lines; the alar extent ten inches and one-fourth; the tail two inches and a half, and reaches an inch beyond the wings. It has the longeſt bill of all the known Finches. [A]

[A] The ſpecific character:—“ It is variegated with duſky and “ yellow, orange below, the tail olive, the head and throat black, “ the collar bay.”

The CHINESE FINCH.

L'Olivette, Buff.
Fringilla Sinica, Linn.
Fringilla Sinensis, Briss.

The base of the bill, the cheeks, the throat, the fore-part of the neck, and the superior coverts of the tail, are of an olive-green; the upper-part of the head, and of the body, of an olive-brown, with a slight rufous tinge on the back, the rump, and the coverts of the wings next the body; the tail black, edged with white, and tipt with whitish; the breast and the belly rufous, mixed with yellow; the inferior coverts of the tail and of the wings, of a fine yellow; the bill and the legs yellowish. It is nearly of the size of a Linnet. The female has the colours, as usual, more dilute.

Total length five inches; the bill six lines; the legs six lines and a half; mid-toe seven lines; alar extent eight inches and one-third; the tail twenty-one lines, forked, and projecting only five or six lines beyond the wings. [A]

[A] The specific character: — "Olive-rufous, below brick-"coloured, the quills of the wings and tail yellowish at the base." Brisson describes it as "dusky-olive, below tawny-yellow; the "fore-part of the head, and the lower part of the neck green-"olive; the first half of the tail-quills yellowish, the other half "black; the tips of the wings whitish."

The EUSTACHIAN FINCH.

*Le Pinſon Jaune & Rouge**, Buff.
Fringilla Euſtachii, Gmel.
Fringilla Inſulæ St. Euſtachii, Briſſ. and Klein.
Paſſer Africanus Eximius, Seba.

Yellow predominates on the throat, the neck, the head, and all the upper-part of the body; on all the extremities, viz. the bill, the legs, the wings, and the tail: theſe two colours meeting together form a beautiful orange on the breaſt, and on all the lower-part of the body. On each ſide of the head there is a blue ſpot immediately below the eye.

Seba ſays that this bird was ſent from the iſland of St. Euſtatius, and he calls it *the African Finch*; probably becauſe this author knew an iſland of St. Euſtachius in Africa very different from that which commonly goes under that name, which is one of the Little Antilles. It is nearly of the ſize of the Chaffinch.

Total length five inches and a half; the bill ſix lines; the legs ſix lines and a half; the mid-toe ſeven lines; the tail twenty-one lines, and extends about ten lines beyond the wings. [A]

* *i. e.* The Yellow and Red Finch.

[A] Specific character:—"Yellow, gold-colour below, with "a cœrulean ſpot below the eyes, the wings and tail red." Briſſon's deſcription is preciſely the ſame.

M

The

The VARIEGATED FINCH.

La Touite, Buff.
Fringilla Variegata, Gmel.
Fringilla varia Novæ Hiſpaniæ, Briſſ.

Seba gives this bird the name of Twite, which it received in New Spain, and which ſeems borrowed from its cry.

This charming bird has its head of a light red, mixed with purple; the breaſt of two ſorts of yellow; the bill yellow; the legs red; all the reſt variegated with red, white, yellow, and blue; laſtly, the wings and tail edged with white. It is nearly of the ſize of the Common Chaffinch.

Total length, five inches and two-thirds; the bill ſix lines and a half; the legs eight lines; the mid-toe ſeven lines and a half; the tail two inches, and it ſtretches eleven lines beyond the wings. [A]

[A] Specific character:—"Variegated with red, yellow, cœrulean, and white, the breaſt clouded with yellowiſh, the tail-quills with a white margin." Briſſon deſcribes it, "variegated like marble, with red, yellow, cœrulean, and white, the head tinged with a faint red mixed with purple; the breaſt whitiſh, ſhaded with deep yellow; the tail-quills edged with white."

The FRIZZLED FINCH.

Le Pinſon Friſé, Buff.
Fringilla Criſpa, Linn. Gmel. and Briff.
The Black and yellow Frizzled Sparrow, Edw.
In Portugueſe, Beco de Prata.

This bird owes its name to the frizzled feathers on its bill and back. Its bill is white; its head and neck black, as if it were a hood of that colour; the upper-part of the body, including the quills of the tail and of the wings, brown olive; the under-part of the body yellow; the legs deep brown.

As this bird came from Portugal, it is preſumed that it was ſent from the principal ſettlements of that nation, viz. the kingdom of Angola in Africa, or from Brazil.

It is nearly of the ſize of the Common Chaffinch.

Total length five inches and a half; the bill five or ſix lines; the tail is compoſed of twelve equal quills, and extends twelve or thirteen lines beyond the wings. [A]

[A] Specific character:—"Olive, yellowiſh below, the head "black, with many reflected feathers." Briſſon deſcribes it, "robed in frizzled feathers, dull olive above, yellowiſh below; the "head and neck black; the tail-quills of a faint olive; the bill "white."

The

The COLLARED FINCH.

Le Pinson a double collier, Buff.
Fringilla Indica, Gmel.
Fringilla Torquata Indica, Briff.
In Portuguefe, *Collherinho*.

This bird has two half-collars, the one before and the other behind; the firft is black, and is the lower of the two, the other is white; the breaft alfo, and all the under-part of the body is ftained with ruft-colour; the throat, the ring of the bill and eyes of a pure white; the head black; all the upper-part of the body cinereous brown, which grows lighter on the fuperior coverts of the tail; the great quills of the wings black; the middle ones and the fuperior coverts black, edged with a gloffy reddifh brown; the bill black, and the legs brown. Briffon fays that it is a native of India. It is as large as the Chaffinch.

Total length about five inches; the bill fix lines; the tail twenty lines; it confifts of twelve equal quills, and projects ten lines beyond the wings. [A]

[A] The Specific character:—" Cinereous dufky rufous-white " below; the bill, the head, the ftripe on the throat, the quills of " the wings and their coverts, the root of the bill, the orbits, " and the upper-part of the neck, white." Thus defcribed by Briffon: " Cinereous dufky above, white below, ftained with tawny; " the head and tranfverfe ftripe on the lower-part of the neck " black; the fmall feathers at the bafe of the bill, the fpace " about the eyes and the throat white; the collar whitifh; the " quills of the wings black, the fmaller ones edged with rufous; " the tail-quills cinereous-dufky."

The MARYGOLD GROSBEAK *.

Le Noir-Souci, Buff.
Loxia Bonarienſis, Gmel.

I have formed an appellation for this new ſpecies from the two principal colours of its plumage: the throat, the fore-part of the neck, and the breaſt are marygold (*ſouci*); the under-part of the body blackiſh (*noiratre*); the quills of the wings and of the tail alſo blackiſh, edged exteriorly with blue; the head and the upper-part of the neck of the ſame colour; the belly and the inferior coverts of the tail ſulphur yellow; the bill blackiſh, ſhort, ſtrong, and convex; the inferior mandible lighter coloured; the noſtrils round, placed in the baſe of the bill, and perforated; the tongue ſemi-cartilaginous and forked; the legs reddiſh brown; the mid-toe joined to the outer one by a membrane, as far as the firſt articulation; the outer toe the largeſt, and its nail the ſtrongeſt; the nails are in general ſharp, hooked, and ſcooped.

Theſe birds appear in pairs; and the male and female ſeem to bear a mutual and faithful attachment: they frequent the cultivated fields and gardens, and live on herbs and ſeeds.

* Specific character:—" Blackiſh, yellowiſh below, the head " and upper-part of the neck cœrulean, the neck and breaſt " tawny."

Commerſon,

Commerſon, who firſt introduced them to our acquaintance, and who obſerved them at Buenos-Ayres in the month of September, aſſigns their rank between the Finches and the Groſbeaks. He ſays they are of the ſize of a Sparrow.

Total length ſeven inches; the bill ſeven lines; alar extent eleven inches and a half; the tail thirty-three lines, and conſiſts of twelve equal quills; the wings have ſeventeen quills, and the ſecond and third are the longeſt of all.

The WIDOWS.

Les Veuves, Buff.

All the ſpecies of Widows are inhabitants of Africa; but they are not entirely confined to that region, for they occur in Aſia, and even in the Philippine Iſlands. They all have a conical bill of ſufficient ſtrength to break the ſeeds on which they feed: they all are diſtinguiſhed by a long tail, or rather by long feathers, which in moſt of the males accompany the true tail, and are inſerted above or below its origin: laſtly, all, or nearly all of them, are ſubject to two annual moultings, the interval between which correſponds to the rainy ſeaſon, and laſts ſix or eight months, during which the males loſe not only their long tail, but their rich colour and pleaſant warble *; and it is not before the return of ſpring that they recover the attributes or ornaments of their ſex.

The females undergo the ſame moultings, but not only is the change leſs perceptible in them, but the colours of their plumage are not ſo much affected.

* The melody of their ſong is one of the reaſons that induces Edwards to claſs them with the Finches rather than with the Sparrows.

The period of the first moulting in the young males must evidently depend on the time of their birth: those of the earliest hatch assume their long tail in May; but those hatched latest in the season, do not assume it till September or October.

Travellers assert that the Widows construct their nest with cotton, and that they divide it into two stories *, the upper being destined for the male, and the under for the female. It is possible to ascertain this circumstance in Europe, and even in France, where by a careful attention the Widows may be made to lay and hatch, as is successfully practised in Holland.

These birds are lively and volatile, and are constantly raising and dropping the tail: they are very fond of bathing, not at all subject to diseases, and live twelve or fifteen years. They are fed with a mixture of spikenard and millet; and by way of cooling, they have leaves of succory.

It is somewhat odd that the name of Widows, by which they are now generally known, and which seems to be very applicable to both, because of the black that predominates in their plumage, and because of their train at the tail, owed its origin to a mere mistake. The Portu-

* *Vide* Kolben's description of the Cape of Good Hope. It appears very probable, that the changeable-plumaged Goldfinches, of which he speaks, are really Widow-birds.

guese called them at first *Birds of Whidha* (that is, of Juida), because they are very common on that coast of Africa; and foreigners were deceived by the similarity between that word and the name of Widow in the Portuguese tongue *.

We shall here treat of eight species of Widows; viz. the five already known and described by Brisson; two new ones which are already distinguished by a beautiful red spot on the wing, and another on the breast: lastly, to these I shall add the Bird which Brisson calls *the Long-tailed Linnet*, which, were it only for the long tail, I should rather range with the Widows than with the Linnets.

* Edwards was led into this mistake, which he afterwards discovered.

The GOLD-COLLARED WIDOW.

Emberiza Paradisæa, Linn. Gmel. Borov.
Vidua, Briss.
Passer Indicus Macrourus alius, Ray, Will. Klein.
The Red-breasted Long-tailed Finch, Edw.
The Whidah Bunting, Lath.

The neck of this bird is covered by a broad half-collar, of a fine yellow gold colour; the belly and thighs are white; the abdomen and the coverts of the under-part of the tail blackish; the head, throat, fore-part of the neck, back,

back, wings, and tail, black. The tail is formed as in other birds; it conſiſts of twelve quills nearly equal, and covered by four long feathers, which riſe alſo from the rump, but ſomewhat higher; the two longeſt are about thirteen inches, and are black, like thoſe of the tail, and appear waved, and as it were clouded; a little arched like thoſe of the cock; their breadth, which is nine lines near the rump, is reduced to three near the extremity: the ſhorteſt are incloſed between the two longeſt, and are only half as long, but they are twice as broad, and end in a ſlender ſilky filament, more than an inch long.

Theſe four feathers have their planes in a vertical ſituation, and are bent downwards; they drop every year in the firſt moulting about the beginning of November, and at this period their plumage ſuffers a total change, and becomes like that of the Brambling. It is now variegated on the head with white and black; the breaſt, the back, the ſuperior coverts of the wings, dirty orange, ſprinkled with blackiſh; the feathers of the tail and the wings of a very deep brown; the belly, and all the reſt of the under-part of the body, white:—Such is its winter garb, which it retains till the vernal ſeaſon, when it undergoes a ſecond moulting as complete as the former, but happier in its effects, for it reſtores the fine colours, the long feathers, and all the decorations; and before the beginning of July the bird has experienced a total renovation. The colour of its eyes, of its bill, and of its legs, never vary: the eyes are chesnut; the bill lead-colour; and the legs flesh-colour.

The young females are nearly of the colour of the males in moult; but at the end of three years, their plumage has become brown, almost black, and changes no more.

These birds are common in the kingdom of Angola, on the western coast of Africa: some have also been sent from Mozambique, a small island on the eastern coast of the same continent, and which differed little from the former. The subject which Edwards figured lived four years in London.

Total length fifteen inches; length measured from the tip of the bill to the end of the nails four inches and a half; the bill four lines and a half; the clear alar extent nine inches; the false tail thirteen inches; the true tail twelve lines, and projecting about an inch beyond the wings. [A]

[A] Specific character of the *Emberiza Paradisæa*:—" Dusky, " the breast red, the four intermediate quills of the tail long and " pointed; two very long, the bill black." Thus described by Brisson: " *In Summer*, glossy black above, tawny white below; the " upper-part of the neck tawny; the breast glossy chesnut, the " tail-quills black; the two intermediate ones long, each pro- " jecting beyond that adjacent; the legs flesh-coloured." " *In* " *Winter*, reddish chesnut above, variegated with dusky spots, and " white below; the head variegated with white and black stripes; " the tail quills dusky blackish, the outer-edges reddish chesnut; " the legs flesh-coloured."

N.º 92

THE SHAFT TAILED BUNTING

The SHAFT-TAILED WIDOW.

La Veuve à quatre brins *, Buff.
Emberiza Regia, Linn. and Gmel.
Vidua Riparia Africana, Briff.
The Shaft-tailed Bunting, Lath.

This bird has the fame two moultings as the preceding, and they are attended with fimilar effects. Its bill and legs are red; the head and all the upper-part of the body black; the throat, the fore-part of the neck, the breaft, and all the lower-part blufh-coloured, but which is brighter on the neck than on the breaft, and extending behind the neck, it forms an half collar, which is broader the lower the black hood defcends from the head. All the feathers of the tail are blackifh, but the four middle ones are four or five times longer than thofe of the fide, and the two middle ones are the longeft of all. In moulting, the male becomes like the Linnet, only it is of a lighter gray. The female is brown, and has not the long feathers in the tail.

This bird is rather fmaller than a Canary; feveral of them are living at Paris, and were all brought from the coafts of Africa.

The average meafures are:—total length twelve or thirteen inches: that from the tip of

* i. e. The Widow with four filaments. It is alfo called *Silk-tail.*

the

the bill to the end of the nails four or five inches; the bill four or five lines; the alar extent eight or nine inches; the two mid-feathers of the tail nine or eleven inches; the two next eight or ten inches; the lateral ones twenty to twenty-three lines. [A]

[A] Specific character:—"The four long intermediate feathers "of the tail equal, and bearded only at their insertion, the bill "red." Brisson describes it, "black above; the neck tawny, "variegated above with black spots; the tail feathers blackish, "the four long intermediate ones furnished with plumules only at "the origin; the bill and legs red."

The DOMINICAN WIDOW.

Emberiza Serena, Linn. and Gmel.
Vidua Minor, Briss.
The Dominican Bunting, Lath.

If length of tail be the distinguishing character of the Widow birds, this is the least entitled to that appellation; for the longest quills of its tail scarcely exceed four inches. It has received the name of *Dominican*, on account of its black and white plumage; all the upper-part of the body is variegated with these two colours; the rump, and the superior coverts of the tail, are mottled with dirty white and blackish; the upper-part of the head of a white-reddish, encircled with black; the throat, the fore-part of the neck, and the breast, of the same white, which also extends behind, and forms a half-collar

collar on the poſterior ſurface of the neck. The belly has none of the rufous tinge. The bill is red, and the legs gray.

This ſpecies undergoes two moultings annually, like the preceding; in the interval the male is diveſted of its long tail, and its white is dirtier. The female never has theſe long feathers of the tail, and its plumage is conſtantly of an almoſt uniform brown.

Length to the end of the tail, ſix inches and one fourth; to the end of the nails, four inches; the bill four lines and a half; the legs ſeven lines; the mid-toe ſeven lines and a half; the alar extent ſeven inches and a half; the middle feathers of the tail project about two inches and one fourth beyond the lateral ones, which are notched, and three inches and one fourth beyond the wings. [A]

[A] Specific character:—"With a black cap, the crown red, "the tail wedge-ſhaped, the two intermediate quills of the tail "longeſt, the bill red." Thus deſcribed by Brisson:—"Black "above, the edges of the feathers rufous, below white verging "to tawny, the crown rufous, the collar white-tawny, the tail-"quills black, the two intermediate ones longeſt, the three next "white at their origin, the two outermoſt tawny on their exterior "edges, and white on their interior; the bill red." Commerſon ſuſpected that a certain bird of a bluiſh-black which he ſaw in the iſle of Bourbon, where it was called *Brenoud*, is nothing but this ſame Widow in moult; and he thence concluded, that when the male moulted its plumage it was more uniform. But this would apply better to the female than to the male; and yet there is a wide difference between bluiſh-black, which is the colour of the Brenoud, and uniform brown, which is that of the female Dominican. This Brenoud reſembles more the Great Widow.

The

The GREAT WIDOW.

Emberiza Vidua, Linn.
Vidua Major, Briff.
Paſſer Indicus Macrourus Roſtro Miniaceo, Ray, and Will.
The Long-tailed Bunting, Lath.

The mourning garb of this Widow is ſomewhat brightened by the fine red colour of the bill, by a tint of bluiſh green ſpread over all the black, that is, over all the upper ſurface; by two tranſverſe bars, the one white, and the other yellowiſh, with which the wings are decorated; and laſtly, by the whitiſh colour of the lower part of the body, and the lateral quills of the tail. The four long feathers inſerted above the true tail * are black, and ſo are the quills of the wings; they are nine inches long, and very narrow. Aldrovandus adds, that the legs are variegated with black and white; and the nails black, very ſharp and hooked. [A]

* Aldrovandus expreſsly obſerves, that the male has a double tail like the peacock, and that the longer reſts upon the ſhorter. It ſeems odd that Briſſon deſcribes the four feathers of the upper tail as the intermediate ones of the true tail.

[A] Specific character:—" Blackiſh, whitiſh below, four inter-" mediate quills of the tail long and pointed, two of them the" longeſt, the bill red." Thus deſcribed by Brisson:—" Black" above, mixed with a greeniſh ſky-colour; whitiſh below, with" a double tranſverſe ſtripe on the wings, the one white, and the" other ſlightly yellowiſh; the four intermediate tail-quills very" long and black, the four exteriors on each ſide whitiſh; the bill" minium coloured."

The

Nº 93

FIG·1·THE LONG TAILED BUNTING.FIG·2·THE SAME AFTER MOULTING.

The ORANGE-SHOULDERED WIDOW.

La Veuve à Epaulettes, Buff.
Emberiza Longicauda, Gmel.
Loxia Longicauda, Mill.
Cape-Sparrow, Kolb.
The Yellow-ſhouldered Oriole, Brown.
The Orange-ſhouldered Bunting, Lath.

The prevailing colour in the plumage of this bird is gloſſy black; and the only exception is in the wings, where the ſmall coverts are of a fine red, and the middle ones of a pure white, which gives the bird a ſort of epaulettes. The large as well as the ſmall quills of the wings are black, edged with a lighter colour.

This bird is found at the Cape of Good Hope. It has, like all the reſt, a double tail; the lower conſiſts of twelve feathers nearly equal, the upper of ſix, which are of different lengths; the longeſt are thirteen inches, and in all, their plane is vertical.

Total length nineteen or twenty-one inches; the bill eight or nine lines; the legs thirteen lines; the tail thirteen inches. [A]

[A] Specific character:—"Black, the ſhoulders fulvous, bordered with white, the tail-quills long, and the ſix intermediate ones project beyond the reſt." It is of the bulk of a Thruſh.

The

The SPECKLED WIDOW.

La Veuve Mouchetée, Buff.
Emberiza Principalis, Linn. and Gmel.
Vidua Angolensis, Briss.
Long-tailed Sparrow, Edw.
The Variegated Bunting, Lath.

All the upper-part is speckled with black on an orange ground; the quills of the wing and its great coverts are black, edged with orange; the breast is of a lighter orange, without speckles; the small coverts of the wing are white, and form a broad transverse bar of that colour, which predominates in all the lower-part of the body; the bill is of a lively red, and the legs flesh-coloured.

The four long feathers' are of a deep black; they constitute no part of the true tail, as might be supposed, but form a sort of false tail which leans on the first. These long feathers are cast in moulting, but quickly replaced; which is common in most birds, though rather unusual in the Widows. When these feathers have acquired their full length, the two middle ones project five inches and a half beyond the lower tail, and the two others an inch less. The quills of the lower or true tail are of a dull brown; the side ones edged exteriorly with a lighter colour, and marked within with a white spot.

This bird is of the size of the Dominican Widow; its bill is of a bright red, shorter than that

that of the ſparrow, and the legs fleſh-colour-ed. [A]

[A] Specific character:—"Variegated, the breaſt rufous, the "four middle tail-quills very long, the bill and legs red." Thus deſcribed by BRISSON:—"Variegated above with black and "rufous, white below; the breaſt ſlightly rufous; the leſſer ſu-"perior coverts of the wings white; four intermediate tail-quills "very long and black; the four outermoſt on each ſide faintly "duſky, edged exteriorly with a ſlighter duſky, ſpotted interiorly "with white; the bill ſaffron."

M

The FIRE-COLOURED WIDOW.

La Veuve en Feu, Buff.
Emberiza Panayenſis, Gmel.
La Veuve de l'Iſle de Panay, Sonn.
The Panayan Bunting, Lath.

This bird is entirely of a fine gloſſy black, except a ſingle red ſpot on its breaſt, which appears like a burning coal. It has four long equal feathers which are inſerted below the true tail, and extend beyond it more than double its length; they grow narrower by degrees, ſo that they terminate in a point. This bird is found at the Cape of Good Hope, and in the iſland of Panay, one of the Philippines; it is of the ſize of the Gold-collared Widow. Its total length is twelve inches. [A]

[A] Specific character:—"Black, a large ſcarlet ſpot on the "breaſt, the four intermediate quills of the tail pointed, very long, "equal and pendulous."

M

The

The EXTINCT WIDOW.

La Veuve Eteinte, Buff.
Emberiza Psittacea, Linn. and Gmel.
Linaria Brasiliensis Longicauda, Briss. and Klein.
Fringilla Brasiliensis, Seba.
The Psittaceous Bunting, Lath.

Brown-cinereous is the prevailing colour of this bird; but the base of the bill is red, and the wings flesh-colour mixed with yellow. It has two quills triple the length of its body, which are inserted in the rump, and tipt with bay-red. [A]

[A] Specific character:—"Cinereous-dusky, the wings fulvous, two of the tail-quills very long." Thus described by Brisson:—"Dull cinereous-gray; the base of the bill encircled with a reddish ring; the wings variegated with dull cinereous gray, the two intermediate ones longest, scarlet at their origin."

The GRENADIN.

Fringilla Granatina, Gmel.
Granatinus, Briss.
The Red and Blue Brazilian Finch, Edw.
The Brasilian Finch, Lath.

The Portuguese, perceiving probably a resemblance between the plumage of this bird, and the uniform of some of their regiments, have named it *the Oronoco Captain*. Its bill and orbits are bright red; its eyes black; on the sides of

No. 94

FIG 1 THE BRASILIAN FINCH.

FIG. 2. THE FRIZLED BRASILIAN FINCH.

of the head is a large plate of purple almoſt round, whoſe centre lies on the poſterior edge of the eye, and which is interrupted between the eye and the bill by a brown ſpot; the throat and the tail are black *; the quills of the wings brown-gray, edged with light-gray; the hind part of the body, both above and below, is of a blue-violet; all the reſt of the plumage is gilded deep brown; but on the back it is variegated with greeniſh-brown, and this ſame gilded deep brown edges exteriorly the coverts of the wings. The legs are of a dull fleſh-colour. In ſome individuals the baſe of the upper-mandible is encircled by a purple zone.

This bird is found in Brazil. Its motions are lively, and its ſong agreeable. It has the long bill of our Goldfinch †, but differs by its extended tapered tail.

The female is of the ſame ſize with the male; its bill red; a little purple under the eyes; the throat, and the under-ſide of the body, pale-fulvous; the top of the head of a deeper fulvous; the back brown-gray; the wings brown; the tail blackiſh; the ſuperior coverts blue, as in the male; the inferior coverts, and the lower belly, whitiſh.

Total length five inches and one fourth; the bill five lines; the tail two inches and a half,

* In ſome ſubjects the throat is of a greeniſh-brown.

† Edwards found the length of the bill to vary in different individuals.

composed of twelve tapered quills, the longest exceeding the shortest by seventeen lines, and the extremity of the wings by two inches; the tarsus seventeen lines; the hind nail the strongest of all. In the wings the fourth and fifth quills are the longest. [A]

[A] Specific character:—"Its tail is wedge-shaped, its body "tawny, its bill red; its temples, its rump, and its lower belly, "violet."

N.º 95

THE GREENFINCH

The GREENFINCH*.

Le Verdier, Buff.
Loxia Chloris, Linn. and Gmel.
Chloris, Aldrov. Gefner, Ray, Sibb. &c.
The Neighing Finch, Charleton.

THIS bird muft not be confounded with the Yellow Bunting *(Bruant)*, though in many provinces it bears the fame name; for, not to mention other diftinctions, it wants the offeous tubercle in the palate.

The Greenfinch paffes the winter in the woods, and fhelters itfelf from the inclemency of the feafon in the ever-green trees, and even in elms and branchy oaks which retain their withered leaves.

In fpring it makes its neft in the fame trees, and fometimes in bufhes: this neft is larger, and almoft as neatly formed as that of the Chaffinch; it confifts of dry herbs and mofs, lined with hair, wool, and feathers: fometimes it places it in the chinks of the branches, which

* It is called in Germany, *Gruenling*, *Gruenfinck*, *Kuttvogel*, *Tutter*, *Rapp-Finck*, *Hirfs-Finck*, *Hirfsvogel*, *Welfcher-Henffling*, *Kirfch-Finck*; in Italy, *Verdon*, *Verderro*, *Verdmontan*, *Zaranto*, *Caranto*, *Toranto*, *Frinfor*; in Portugal, *Verdelham*; in Savoy, *Verdeyre*; in Illyria, *Zeglolka*; in Bohemia, *Schwonetz*; in Pruffia, *Gruener-Henffling*, *Schwontzke*; in Poland, *Dzwonieck*, *Konopka*; in Sweden, *Swenfka*.

it

it even widens with its bill; it alſo conſtructs near the ſpot a little magazine for proviſions*.

The female lays five or ſix eggs, ſpotted at the large end with brown red on a white greeniſh ground. She ſits aſſiduouſly, and ſtill continues on her eggs though a perſon approaches pretty near; ſo that ſhe is often caught with her young: at all other times ſhe is ſhy and timid. The male ſeems to take much intereſt in the concerns of his future family; he relieves his mate in hatching; wheels round the tree where his hopes are lodged, makes ſudden ſprings, and again ſinks back, flapping his wings, and warbling joyous notes†. At his return to the country, and at his departure, he utters a ſingular cry, conſiſting of two ſounds. The warble is ſaid to be improved in the croſs breed between the Greenfinch and the Canary.

The Greenfinches are gentle, and eaſily tamed; they learn to articulate a few words; and no bird ſo ſoon becomes expert at the little manœuvre of drawing up the cup‡. They eat from the finger of their maſter, and anſwer his call, &c. In autumn they join other ſpecies, to roam in the fields; they live upon juniper berries in winter; they crop the buds of trees, and

* We owe theſe laſt facts, and ſome others, to M. *Guys*.

† They are kept in a cage, becauſe they ſing pleaſantly. BELON. Guys adds, that the warble of the female is even ſuperior to that of the male, which would be ſingular in birds.

‡ *De la galere*, alluding to the labour of a galley-ſlave.

parti-

particularly thoſe of the bog-willow; they feed in ſummer on all ſorts of ſeeds, and eſpecially thoſe of hemp; they alſo eat caterpillars, ants, and graſshoppers.

The name alone denotes that the predominant colour of the plumage is green; but the tinge is not pure; there is a gray-brown caſt on the upper-part of the body and on the flanks, with an admixture of yellow on the throat and breaſt; yellow is ſpread over the top of the belly, the inferior coverts of the tail and wings, and on the rump; it edges the largeſt quills of the wings, and alſo the lateral quills of the tail: all theſe are blackiſh, and moſt of them bordered with white on the inſide; the lower belly is alſo white, and the legs reddiſh-brown.

The female has more brown; her belly is entirely white, and the inferior coverts of her tail are mingled with white, brown, and yellow.

The bill is fleſh-coloured, ſhaped like a cone, and ſimilar to that of the Groſbeak, but ſmaller; its upper edges are ſlightly ſcalloped near the point, and receive thoſe of the lower mandible, which are ſomewhat *re-entrant*. The bird weighs rather more than an ounce, and is nearly of the bulk of the Houſe-ſparrow.

Total length five inches and a half; the bill ſix lines and a half; the alar extent nine inches; the tail twenty-three lines, ſomewhat forked, ſtretching beyond the wings ten or eleven lines. Theſe birds have a gall-bladder, a muſcular gizzard covered with a loose membrane, and a pretty large craw.

Some pretend that there are Greenfinches of three different sizes; but this is not sufficiently ascertained, and probably such variations are only accidental, resulting from age, from food, from climate, and from other like circumstances. [A]

[A] Specific character:—" Yellow-green, the primary wing-" quills yellowish before, the four lateral tail-quills yellowish at " their base." The Greenfinch is very common in Great Britain, and usually nestles in the hedges.

M

The PAINTED BUNTING*.

Le Pape, Buff.
Emberiza-Ciris, Linn. Gmel.
Fringilla Mariposa, Scop. Ann.
Fringilla Tricolor, Klein.
Chloris Ludoviciana, *Papa*, Briss.
China Bulfinch, Alb.
The Painted Finch, Edw. and Catesby.

This bird has its name *(Pope)* from the colours of its plumage, and especially from a sort

* Specific character:—" The head cœrulean, the lower bill " fulvous, the back green, the quills dusky-green." Thus described by Brisson: " Above green, inclining to yellow, below red; " the head and the upper-part of the neck cœrulean-violet; the " rump red; the tail-quills dusky, both sides of the two inter-" mediate ones varying to red, and the outer surface of the la-" teral ones the same." The Spaniards of Vera Cruz, which it visits in winter, call it *Maripofa Pintada*, or " the Painted Butterfly."

of

N.º 96

FIG. 1. THE CHINA BULFINCH.
FIG. 2. THE BLUE BULFINCH.

of capuchin, which riſes at the baſe of the bill, and extending below the eyes covers the upper and ſide-parts of the head and neck, and in ſome individuals returns under the throat. The fore-part of the neck, all the lower-part of the body, and even the ſuperior coverts of the tail and of the rump, are of a fine red, almoſt fire-coloured; the back variegated with ſoft green and dull olive *; the great quills of the wings and of the tail are of a reddiſh brown, the great coverts of the wings are green; the ſmall ones are of a violet blue, like the capuchin. But Nature requires time to form a plumage ſo charming, nor is it completed before the third year: the young Painted Buntings are brown the firſt year; in the ſecond, their head is of a vivid blue, the reſt of the body greeniſh blue, and the quills of the wings and of the tail are brown, edged with greeniſh blue.

But it is the female chiefly which reſembles the Greenfinch; the upper-part of its body is of a dull green, and all the under-part of a yellowiſh green; the great quills of the wings are brown, edged delicately with green; the middle ones, and alſo the quills of the tail, divided length-wiſe into brown and green.

Theſe birds breed in Carolina on the orange trees, but do not continue there during the

* In the ſubject deſcribed by Cateſby, the green back was terminated with yellow.

winter. Like the Widows, they have two moultings annually, which are earlier or later according to circumſtances. Sometimes they aſſume their winter garb in the end of Auguſt or the beginning of September; in this ſtate the under-part of their body changes from red to yellowiſh. They feed like the Widow-birds, upon millet, Canary-ſeeds, ſuccory, &c. . . . but they are more delicate: however, if they are once ſeaſoned to the climate, they will live eight or ten years. They are found in Louiſiana.

The Hollanders have been able, by care and patience, to breed the Painted Buntings in their country, as they have ſucceeded with the Bengals and Widows; and it is likely that with the ſame attention they might be propagated in every part of Europe —They are rather ſmaller than the Houſe-ſparrow.

Total length five inches and one-third; alar extent ſeven inches and two-thirds; the bill eight lines; the legs eight lines; the middle toe ſeven lines; the tail two inches, and ſtretches thirteen or fourteen lines beyond the wings.

VARIETY

VARIETY of the PAINTED BUNTING.

Bird-fanciers are acquainted with a variety of this ſpecies, which is diſtinguiſhed by the colour of the under-part of its body being yellowiſh; it has only a ſmall red ſpot on the breaſt, which is loſt in moulting; then all the under-part of the body is whitiſh, and the male much reſembles the female. It is probably a variety produced in this climate.

M

The BLUE-FACED BUNTING*.

Le Toupet Bleu, Buff.
Emberiza Cyanopſis, Gmel.
Chloris Javenſis, Briſſ.

This bird reſembles the Painted Bunting ſo cloſely, that if the one had not been ſent from Louiſiana, and the other from Java, I ſhould have regarded them as of the ſame ſpecies. Nay, when we conſider the uncertainty in moſt ac-

* Specific character:—“ Green, the rump and lower-belly “ rufous; the forehead, cheeks, and throat cœrulean; the quills “ of the wings duſky and edged with green; the tail-quills edged “ with red, the intermediate ones green, the reſt duſky.” Thus deſcribed by Briſſon: “ Above green, below rufous; middle of the “ belly red; rump rufous; the forehead, cheeks, and throat “ cœrulean, the wing-quills green, their exterior borders red, “ the lateral ones duſky on the inſide.”

counts

counts of the climate of birds, we may ſtill be diſpoſed to aſſign them to the ſame place. The anterior-part of the head and throat is of a fine blue; the fore-part of the neck of a fainter blue; the middle of the belly red; the breaſt, the flanks, the lower belly, the thighs, the inferior coverts of the tail and of the wings, of a beautiful rufous; the upper-part of the head and neck, the anterior-part of the back, and the ſuperior coverts of the wings, green; the lower part of the back and the rump of a bright rufous; the ſuperior coverts of the tail red; the quills of the wings brown, edged with green; thoſe of the tail the ſame, except the intermediate ones, which are edged with red; the bill is lead-colour; the legs gray:—the bird is ſomewhat ſmaller than the Tree-ſparrow.

Total length four inches; the bill ſix lines; the legs ſix lines and a half; the mid-toe ſeven lines; the alar extent near ſeven inches; the tail thirteen lines, compoſed of twelve quills, and ſtretching ſix or ſeven lines beyond the wings.

The GREEN BUNTING*

Le Parement Bleu, Buff.
Emberiza Viridis, Gmel.
Chloris Indica Minor, Briſſ.

All our information with regard to this bird is derived from Aldrovandus; and that author only deſcribes it from a coloured drawing brought into Italy, by ſome who had viſited Japan, and who preſented it to the Marquis Fachinetto.

All the upper-part is green, and all the lower white; the quills of the tail and of the wings blue, with white borders; the bill of a greeniſh brown, and legs black. Though this bird is rather ſmaller than the Greenfinch, and its bill and legs more ſlender, Aldrovandus was convinced that Ariſtotle himſelf could not heſitate in referring it to that ſpecies: this Briſſon has done, and we have no reaſon to reject the arrangement.

* Specific character:—" Above green, below white; the " wing-quills and thoſe of the tail cœrulean."—Briſſon gives the ſame deſcription, only he adds, " that the wing and tail-quills " have white ſhafts."

The YELLOW FINCH*.

Le Vert-Brunet, Buff.
Fringilla Butyracea, Linn.
Chloris Indica, Briss.

Its bill and legs are brown; the upper-part of the head and neck, the back, the tail, and wings of a very deep green brown; the rump, the throat, and all the lower-part yellow; the sides of the head varied with the two colours, so that the yellow descends a little on the sides of the neck.

The Indian Greenfinch of Edwards may be regarded as a variety of this species; for all the upper-part is of a brown green, and the under-part yellow: the only difference being, that the green is not so deep, and extends upon the rump; but the sides of the head have two bars of the same colour, one of which stretches above the eyes, and the other, which is shorter, and of a deeper shade, lies under the first; and that the great quills of the wings are edged with white. The Indian Greenfinch is rather larger than

* Specific character:—" Green, the eye-brows, breast, and " lower-belly yellow, the primary wing-quills white on their " exterior edge." Described thus by Brisson: " Above of an " olive-green, below yellowish; with a stripe on both sides, which " is yellowish above the eyes, olive-green contiguous to them, " and black below them; the quills of the wings olive-green, " the outer-edge of the larger white; the quills of the tail faint " yellow-green."—It is found in India, and at the Cape of Good Hope.

than the Canary Finch, and according to Edwards, its ſong is ſuperior.

Total length four inches and a half; the bill four lines and a half; the tarſus ſix lines and a half; the mid-toe ſeven lines; the tail nineteen lines, ſomewhat forked, and extending nine or ten lines beyond the wings.

M

The BAHAMA FINCH*.

Le Verdinere, Buff.
Fringilla Bicolor, Linn. and Klein.
Chloris Bahamenſis, Briſſ.

Except the head, neck, and breaſt, which are black, all the reſt of the plumage is green; we might call it a Greenfinch with a black cowl. This bird is very common in the woods of the Bahama Iſlands; it ſings perched on the tops of buſhes, and conſtantly repeats the notes of the Chaffinch. It is about the ſize of the Canary Finch.

Total length four inches; the bill four lines and a half; the tail nineteen lines, and extends nine or ten lines beyond the wings.

* Specific character:—" The head and breaſt black; the back, " wings, and tail obſcure greeniſh." Briſſon's deſcription is preciſely the ſame—It is found alſo in Jamaica.

M

The

The GREENISH FINCH.

Le Verderin, Buff.

It has leſs green in its plumage than thoſe of the preceding articles: its bill is ſhorter; its orbits greeniſh-white; all the feathers of the upper-part of the body, including the middle quills of the wings, their coverts, and the quills of the tail, are of a brown-green, edged with a lighter colour; the great quills of the wings black; the throat and all the under-part of the body, as far as the thighs, of a dull rufous, ſpeckled with brown; the lower belly and the inferior coverts of the tail are of a pretty pure white. This bird is found in St. Domingo.

The VARIED GREENFINCH.

Le Verdier ſans Vert, Buff.

This bird has none of the green colour of the Greenfinch, but is cloſely related to it in other reſpects. Its throat is white, the under-part of the body of the ſame colour; the breaſt variegated with brown; the upper-part of the head and body mottled with gray and greeniſh-brown; there is a rufous tinge on the termination of the back and on the ſuperior coverts of the tail; the ſuperior coverts of the wings of a deep rufous;

the

the middle quills edged on the outſide with that colour; the great quills and the great coverts edged with ruſty white, and alſo the lateral quills of the tail; laſtly, the moſt exterior of the latter is terminated by a ſpot of the ſame white, and is ſhorter than the reſt. Of all the quills of the wing, the ſecond and third are the longeſt.

This bird was brought from the Cape of Good Hope by Sonnerat.

Total length ſix inches and one-third; the bill ſix lines; the tarſus ſeven lines; the tail about two inches and a half, and extends ſixteen lines beyond the wings.

The GOLDFINCH*.

Le Chardonneret, Buff.
Fringilla Carduelis, Linn. and Gmel.
Carduelis, Aldrov. Gef. Johnft. Sibb. &c.
Goldfinch, or *Thiftle-finch*, Penn. Rufs. Will. Alb. &c.

BEAUTY of plumage, melody of fong, fagacity, and docility, are united in this charming little bird, which, were it rare, and imported from a foreign country, would be highly prized.

Crimfon red, velvet black, white, and gold yellow, are the chief colours which gliften on its plumage; and the mixture of lighter and deeper tints ftill heightens their luftre. Hence its names in different languages: fome allude to the yellow fpot that decorates its wings †; fome to the red which covers its head and throat ‡; fome to the brilliancy of its colours §; and others, to the

* In Germany it is called *Stieglitz, Diftel-Vogel, Kletter, Truns, Roth-Vogel*; in Holland, *Pitter*; in Bohemia, *Steglick*; in Savoy, *Charderaulat*; in Poland, *Sczigil*; in Sweden, *Stiglitza*; in Italy, *Gardello, Gardellino, Cardelino, Carzerino*; in Spain, *Sirguerito, Siele Colore, Forte Pintacilgo*.

The Latin epithet *Carduelis* is derived from *Carduus*, a thiftle; and the French name *Chardonneret* is formed from *Chardon*, which alfo fignifies a thiftle. It is the Οραυπις of Ariftotle.

† Χρυσομιτρης *(Golden-mitred)*; *aurivittis (with golden fillets)*; *Goldfinch*.

‡ *Roth-Vogel*, Ger. (i. e. *Red-bird.)*

§ Αστηρης. αστρωλινος (from αστηρ, *a ftar.)*

13 effect

No. 97

THE GOLDFINCH

effect of their variety *. When the wings are closed, each appears marked with a train of white points, which are apparent on the dark ground: these are the white specks which terminate all the quills of the wing, except the two or three first. The quills of the tail are of a still deeper black; the six intermediate ones tipt with white, and the two last have on each side on their inner webs an oval white spot, which is conspicuous. But these white points vary in their number and arrangement; and in general the plumage of the Goldfinch is far from being constant †.

The female has less red than the male, and no black at all. The young ones do not assume their fine red till the second year: they are at first stained with dirty obscure colours, and, for that reason, they are called *Grisets;* but the yellow on the wings appears early, and also the

* Ποικιλις, *varied* (variegated).

† Sometimes six intermediate quills of the tail are tipt with white, sometimes eight of these, and sometimes only four, or even two: and the Goldfinches have received appellations accordingly. Nay, the difference observed in their song has been attributed to the number of the small spots. Those which have six feathers tipt with white are said to warble the sweetest; but this assertion is without foundation, for the number is often diminished by moulting, though the song remains invariably the same. Kramer says, that the quills of the tail and of the wings are tipt with white in autumn, and are entirely black in spring: this needs to be qualified. I have beside me at present (6th April), two cock Goldfinches, in which all the wing-quills except the two first and the six intermediate ones of the tail, are tipt with white; and in which are also the oval spots on the inner-side of the two lateral quills of the tail.

 white

white ſpots on the quills of the tail; yet theſe ſpots are of a duller white*.

The males have a well-known and a pleaſant warble. They begin about the 1ſt of March, and continue during the whole of the genial ſeaſon: they chant even in the winter when kept in warm apartments, where they enjoy the temperature of ſpring†. Aldrovandus ranks them the ſecond of the ſinging birds: Daines Barrington admits them only into the ſixth place. They ſeem to have a greater facility in acquiring the ſong of the Wren than that of any other bird‡. This has been experienced both by Salerne and Barrington. The latter indeed ſuppoſes, that this imitation was occaſioned by the early impreſſions made by the notes of that little bird; but we muſt either account in the ſame way for the caſe noticed by Salerne, or admit that there is a certain analogy between the organs of ſound in the Goldfinch and the Wren.

In England, the Goldfinches from Kent are reckoned the beſt ſingers.

* Obſerved before the 15th June. I have alſo remarked, that when the Goldfinches were quite young, their bill was brown, except the point and edges, which were whitiſh and tranſparent; which is the reverſe of their appearance when adult.

† I have two which ſang every day through the whole of this winter, kept in a cloſe chamber, but without any fire. The greateſt cold has not indeed been under eight degrees (fifty degrees Fahrenheit).

‡ *Philoſ. Tranſ.* 1773. Olina ſays, that the young Goldfinches which hear Linnets, Canaries, &c. acquire their ſong. But I know of a young cock Goldfinch and a young hen Linnet that were bred together; and the Goldfinch retained his native warble, while the Linnet adopted and improved it.

Theſe birds and the Chaffinches conſtruct the neateſt and moſt compact neſts. They conſiſt of fine moſs, lichens, liver-wort, ruſhes, ſmall roots, and the down of thiſtles, interwoven with great art, and lined with dry graſs, hair, wool, and down: they generally place them in trees, and particularly on plums and walnuts, and commonly ſelect the weak branches which ſhake the moſt. Sometimes they neſtle in copſes, and ſometimes in thorny buſhes; and it is ſaid that the young Goldfinches bred in ſuch ſituations are of a darker plumage, but more ſprightly, and ſing better than others. Olina makes the ſame remark in regard to thoſe hatched in the month of Auguſt. The female begins to lay about the middle of ſpring, at firſt, five eggs, ſpotted with reddiſh brown near the large end *. If the neſt be demoliſhed, ſhe makes a ſecond or even a third hatch, but the number of eggs diminiſhes each time. I have never found more than four eggs in the neſts brought to me in July, nor more than two in thoſe of September.

Theſe birds are much attached to their young; they feed them with caterpillars and inſects: if both be caught together and ſhut up in the ſame cage, the parents will ſtill continue their care. It is true, that of four young Goldfinches which

* Belon ſays, that the Goldfinches have commonly eight young; but I have never found more than five eggs in about thirty neſts which I have ſeen.

 I treat-

I treated this way, none lived more than a month; but I attribute this to the improper choice of food, and not to the heroic despair which, it is pretended, prompts them to kill their offspring, and thus deliver them from slavery *.

The cock Goldfinch ought to have only one female, and, that their union may be productive, both should be at liberty. It is somewhat singular, that the cock Goldfinch, when shut in a volery, is not so much disposed to pair with his own female as with the hen Canary †, or any other female of a warm temperament.

Sometimes the hen Goldfinch breeds with the cock Canary, but this is unfrequent ‡; and on the other hand, the hen Canary, if separated from the males, will consort with the cock Gold-

* Gerini, and many others. It is even added, that if the young ones be reared in a cage by parents which are suffered to enjoy freedom, these observing after some time the impossibility of rescuing their brood from bondage, will out of compassion poison them with a certain herb. Such fables need no refutation.

† It is said that the Goldfinches intermix with none of a different species; and that the experiment has been made without success in regard to the Linnets. But I confidently affirm, that with proper care we may obtain not only this, but many other combinations: for the Siskins are still more easily reconciled to the society of the Canaries than the Goldfinches, and yet it is said that, in case of rivalship, the Goldfinches are preferred by the hen Canaries.

‡ Father Bougot let a cock and hen Goldfinch into a volery where there was a great number of Canaries of both sexes. The male Canaries copulated with the hen Goldfinch, while the cock Goldfinch had no partner: which proves the ardour of the Canaries.

 finch.

finch *. The female is the firſt to feel the ardour of paſſion, and endeavours, by all alluring arts, and aſſiſted by the ſtill more powerful influence of the genial ſeaſon, to urge her languid paramour to conſummate this phyſical adultery: and yet there muſt be no female of his own ſpecies in the volery. The preliminaries laſt commonly ſix weeks, during which time the hen Canary makes a complete laying of addle eggs, for which her inceſſant ſolicitations have not procured fecundation; what in animals may be termed luſt, is almoſt always ſubordinate to the great end of nature, the reproduction of the ſpecies. Father Bougot, whom I have already quoted with approbation, has obſerved with attention the little manœuvres of the variegated female Canary in ſuch ſituations: ſhe often approached the male Goldfinch, and ſquatted like a common hen, but with more expreſſion, to invite him to the fruition: at firſt he is deaf to her ſolicitations, but the flame ſteals upon him by degrees †: often he begins the embrace, and his reſolution as often fails;

* This circumſtance is eſſential; for Father Bougot aſſures me, that if the female Canaries have a fourth or even a ſixth of their number of males, they will not aſſociate with the cock Goldfinch; and that it is only when neglected by their own ſpecies that they join the ſtranger, in which caſe they always make the firſt advances.

† I have heard it alleged that the Goldfinch was a cold bird; he may be ſo in compariſon of the Canaries, but after he is rouſed, he appears much animated; the male has frequently been obſerved to drop in an epileptic fit, while he chanted rapturouſly in the height of his paſſion.

fails; at each attempt he ſpreads his wings and gives vent to feeble cries. However, after the union is accompliſhed, he proceeds to diſcharge the duties of a parent; he aſſiſts his mate in conſtructing the neſt *, and carries food to her while ſhe is engaged in hatching, or in rearing her young.

Though theſe conſtrained amours will ſometimes ſucceed with a hen Canary and a wild Goldfinch, it is better to raiſe together thoſe intended for breeding, and not to pair them till they are two years old. The offspring reſembles more the father in the ſhape of the bill, in the colours of the head and of the wings, and in ſhort in all the extremities, and the mother in the reſt of the body: they have alſo been remarked to be ſtronger, and live longer; and to have a clearer natural warble, but to be not ſo docile in adopting the notes of our artificial muſic.

Theſe hybrids are ſtill capable of propagating, and when they are paired with the hen Canary, the ſecond generation has a manifeſt analogy to the ſpecies of the Goldfinch †; ſo much does the male influence predominate in the act of generation.

The Goldfinch flies low, but with an even continued motion, like the Linnet; and not by jerks and bounds, like the Sparrow. It is an

* They prefer moſs and dry graſs for the materials.

† Hebert.

active

active and laborious bird; if not employed in pecking the heads of poppies, of hemp, or of thistles, it is constantly busy in carrying backwards and forwards whatever it can find in its cage. One solitary male of this species is sufficient to disturb a whole volery of Canaries; it teazes the females while they are sitting, fights with their males, tears away the nests, and breaks the eggs. We should hardly conceive that birds so lively and petulant could be so gentle and even so docile. They live in harmony with each other, seek each other's society, give marks of regard at all seasons, and seldom quarrel but about their food. They are not so peaceful with other birds; they beat the Canaries and Linnets, but, in their turn, they receive the same treatment from the Titmouse. They have a singular instinct of always chusing to sleep in the highest part of the volery; and we may naturally suppose a ground of quarrel will be afforded, when the other birds will not give place to them.

The docility of the Goldfinches is well known: they can be instructed without much trouble to perform several movements with accuracy, to fire a cracker, and to draw up small cups containing their food and drink; but for this last purpose they must be *clothed*. This clothing consists of a small belt of soft-leather two lines broad, with four holes through which the feet and wings are passed, and the ends joining under

der the belly, are held by a ring which ſupports the chain of the cup. In ſolitude it delights to view its image in the mirror, fancying it ſees another of its own ſpecies; and this attachment to ſociety ſeems to equal the cravings of nature; for often it is obſerved to pick up the hemp ſeed, grain by grain, and advance to eat at the mirror, imagining, no doubt, that it feeds in company.

To ſucceed in breeding Goldfinches, they ought to be ſeparated and raiſed ſingly, or at leaſt each with the female with which it is intended to be paired.

The younger Madame Daubenton educated a whole hatch; the young Goldfinches became tame at a certain age, and afterwards relapſed into the ſame ſavage ſtate in which they would be found if bred by their parents in the field: they renounce the refinements of man to enjoy the ſociety of their fellows.—But this is not the only inconvenience of training them together; they acquire an affection for each other, and when ſeparated, to pair with a female Canary, they are languid in their amours, are affected by the tender remembrance of former friendſhips, and commonly die of melancholy *.

* Of five Goldfinches bred in the volery of this lady, and paired with hen Canaries, three remained inactive: the other two copulated, but broke the eggs, and died ſoon afterwards.

The

The Goldfinches begin to aſſemble in autumn, and during that ſeaſon they are caught among the birds of paſſage which pillage our gardens. Their natural vivacity precipitates them into the ſnares; but, to ſucceed well, it is neceſſary to have a male that has been accuſtomed to ſing. They are not caught by calls; and they elude the bird of prey by taking ſhelter among the buſhes. In winter they fly in numerous flocks, ſo that ſeven or eight may be killed at a ſhot; they approach the highways near which are thiſtles and wild ſuccory; they ſhake off the ſnow to obtain the ſeeds and the caterpillars. In Provence they lodge in great numbers among the almond-trees; when the cold is intenſe they ſeek the cover of thick buſhes, and always near their proper food. Thoſe kept in the cage are commonly fed with hemp-ſeed*. They live to a great age; Geſner ſaw one at Mayence which was twenty-three years old; they were obliged once a-week to ſcrape its nails and bill, that it might drink, eat, and ſit upon its bar; its common food was poppy-ſeeds; its feathers were all turned white; it could not fly, but remained in whatever ſituation it was placed. In the country where I reſide it ſometimes lives ſixteen or eighteen years.

* Though it is true in general that the granivorous birds live upon ſeeds, they alſo eat caterpillars, ſmall grubs and inſects, and even feed their young with the laſt; they alſo devour with great avidity ſmall rags of boiled veal; but ſuch as are reared prefer in the end hemp and rape-ſeed to every other aliment.

They

They are ſubject to epilepſy, as I have already obſerved *, and to melting of the fat; and the moulting often proves fatal to them.

Their tongue is parted at the tip into ſmall filaments; the bill long, the edges of the lower mandible fitted into the upper; the noſtrils covered with ſmall black feathers; the outer toe connected to the middle one as far as the firſt joint; the inteſtinal canal a foot long; ſlight traces of a *cæcum*; a gall bladder; and a muſcular gizzard.

Total length of the bird five inches and ſome lines; the bill ſix lines; the alar extent eight or nine inches; the tail two inches, and conſiſts of twelve quills; it is a little forked, and projects ten or twelve lines beyond the wings †. [A]

* Owing, it is ſaid, to a long ſlender worm which creeps under the fleſh in the thigh, and ſometimes pierces through the ſkin, but which the bird eradicates with its bill. I doubt not the exiſtence of theſe worms, which Friſch mentions; but I ſuſpect that they are not the cauſe of the epilepſy.

† The young Goldfinches are not ſo long in proportion.

[A] Specific character of the Goldfinch, *Fringilla Carduelis*:—"Its wing-quills yellow before, the outermoſt ſpotleſs; the two "outmoſt tail-quills white in the middle, and the reſt at the tip." Thus deſcribed by BRISSON:—"It is tawny-brown; the fore-"part of its head, and its throat, red; its wing-quills white at the "tip, the firſt half yellow exteriorly; the tail-quills black, the ſix "intermediate ones white at the tip, the two outmoſt on either "ſide ſpotted interiorly with white." The ſhoulders of the male are black, thoſe of the female cinereous; in the former the feathers at the baſe of the bill are black, in the latter they are brown. The young bird is gray-headed, and thence named by the bird-catchers *a gray pate*. The egg is pearly, with ſcattered bloody and blackiſh ſpots.

VARIETIES of the GOLDFINCH.

Though the Goldfinch when kept in the cage does not ſo ſoon loſe its red tinge as the Linnet, yet, like all the domeſticated birds, it is ſubject to frequent and material alterations in its plumage. I have already noticed the varieties of age and ſex, and alſo the numerous differences that occur between individuals, in regard to the number and diſtribution of the ſmall white ſpots of the tail and of the wings, and alſo with reſpect to the lighter or darker brown caſt of the plumage. I ſhall here conſider only the principal varieties which I have examined myſelf, or which have been deſcribed by others, and which appear to me as derived from accidental cauſes.

I. The YELLOW-BREASTED GOLDFINCH. It is not unfrequent to ſee Goldfinches which have the ſides of their breaſt yellow, and the ring on their bill and their wing-quills of a lighter black. It has been ſuppoſed that they ſing better than the others; it is certain that in the female the ſides of the breaſt are yellow as in the male.

II. The GOLDFINCH WITH WHITE EYEBROWS AND FOREHEAD *. What is commonly red about the bill, and the eyes, in birds of

* *Fringilla Carduelis*, var. 1. Linn.
Carduelis Leucocephalos, Briſſ.

of this kind, was white in the preſent. Aldrovandus mentions no other difference. I have ſeen a Goldfinch in which the part of the head uſually black was white.

III. The GOLDFINCH WHOSE HEAD IS STRIPED WITH RED AND YELLOW *. It was found in America, but probably carried thither. I have remarked in ſeveral Goldfinches, that the red of the head and throat was variegated with ſhades of yellow, and alſo with the blackiſh colour of the ground of the feathers, which in ſome parts gave a dark caſt to the brilliant colours of the ſurface.

IV. The BLACK-HOODED GOLDFINCH †. The red peculiar to the Goldfinch is alſo found in this variety, but in ſmall ſpots ſprinkled on the forehead. Its wings and tail are alſo as uſual; but the back and breaſt are of a yellowiſh brown; the belly and thighs of a pure white, the iris yellowiſh, and the bill and legs fleſh coloured.

Albin was informed by a *perſon of credit*, that this individual was bred by a female Goldfinch with a male Lark. But a ſingle teſtimony is not ſufficient to ſupport ſuch an aſſertion; Al-

* *Fringilla Carduelis*, var. 2. Linn.
Carduelis Capite Striato, Briſſ.

† *Fringilla Carduelis*, var. 3. Linn.
Carduelis Melanocephalos, Briſſ.
The Swallow Goldfinch, Alb. and Lath.

bin

bin adds by way of confirmation, that it bore ſome reſemblance to the Lark in its ſong and its habits.

V. The WHITISH GOLDFINCH *. If we except the upper-part of the head and the neck, which were of a fine red, as in the Common Goldfinch, the tail which was aſh-brown, the wings which were the ſame, with a bar of dirty yellow, the plumage of this bird was whitiſh.

VI. The WHITE GOLDFINCH †. That of Aldrovandus was, like the common kind, marked with red on the head, and ſome of the wing-feathers were edged with yellow; all the reſt were white.

That of Abbé Aubry had a yellow tinge on the ſuperior coverts of the wings, ſome of the middle quills black through their outer half, and tipt with white; the legs and nails white, the bill of the ſame colour, but blackiſh near the end.

I ſaw one at Baron de Goula's, of which the throat and forehead were of a faint red, the reſt of the head blackiſh; all the under-part of the body white, ſlightly ſtained with aſh-gray, but purer immediately under the red of the throat, and which roſe as far as the blackiſh head; its wings

* *Fringilla Carduelis*, var. 4. Linn.
Carduelis Albida, Briſſ.

† *Fringilla Carduelis*, var. 5. Linn.
Carduelis Candida, Briſſ.

wings yellow, as in the Common Goldfinch; the ſuperior coverts olive; the reſt of the wings white, with a cinereous caſt on the quills neareſt the body; the tail nearly of the ſame white; the bill of a roſe-white, and very long; the legs fleſh-coloured. This variety is the more remarkable, being the production of nature; it was caught full grown in the fields.

Geſner was told that Goldfinches are found entirely white in the country of the Griſons.

VII. The Black Goldfinch*. Several have been ſeen of that colour. That of Aſpernacz mentioned by Anderſon, grew quite black, after being long kept in a cage.—An inſtance preciſely the ſame happened in the town where I live.

In the one deſcribed by Briſſon, four quills of the wings, from the fourth to the ſeventh incluſive, were edged with a fine ſulphur-colour on the outſide, and white on the inſide; the interior of the middle quills was alſo white, and one of them was tipt with the ſame colour; laſtly, the bill, the legs, and the nails, were whitiſh.—But it is obvious that deſcription preſents only one view of a fleeting exiſtence; the object of a general hiſtory is to trace the gradation of appearances, and to connect the individual with the ſpecies.

* *Fringilla Carduelis*, var. 6. Linn.
Carduelis Nigra, Briſſ.

There

There are at present two Black Goldfinches at Beaune, of which I have obtained some information. They are two males, the one four years old, and the other of a greater age; each has undergone three moultings, and has as often recovered its beautiful colours: at the fourth moulting both have become of a pure glossy black; they have retained this colour about eight months, but it appears not more constant than the first, for now (March the twenty-fifth) they begin to perceive gray on the belly of one of these birds, red on the head, rufous on the back, yellow on the quills of the wings, and white at their tips and on the bill. It would be curious to discover how these changes of colours are effected by the food, the air, the temperature, &c. It is known that the Goldfinch which Klein electrified, lost entirely in the course of six months not only the red on its head, but the fine lemon spot on its wings.

VIII. The ORANGE-HEADED BLACK GOLDFINCH *. Aldrovandus found this bird to be so different from the Common Goldfinch, that he regarded it as of another species belonging to the same genus; it was as large as the Chaffinch; its eyes were proportionally larger; the upper-part of its body was blackish, the head of

* *Fringilla Carduelis*, var. 7. Linn.
Carduelis Nigra Icterocephalos, Briss.
Carduelì Congener, Ray, and Will.

the

the ſame colour, except that the anterior part near the bill was encircled by a ring of vivid orange; the breaſt, and the ſuperior coverts of the wings, of a greeniſh black; the outer edge of the wing-quills the ſame, with a bar of faint yellow, and not of a fine lemon, as in the Goldfinch; the reſt of the quills black, variegated with white; thoſe of the tail black, the outermoſt one edged interiorly with white; the belly cinereous brown.

This variety of colour was not owing to the effects of confinement. The bird was caught near Ferrara, and ſent to Aldrovandus.

IX. The Hybrid Goldfinch*. Many of theſe have been obſerved, and it would be tedious and unneceſſary to deſcribe them all. We may aſſert in general that, as in the mule quadrupeds, they reſemble the father moſt in the extremities, and the mother in the reſt of the body. But theſe are not real Hybrids, for they are bred between congenerous birds, ſuch as Canaries, Goldfinches, Greenfinches, Siſkins, Yellow Buntings, and Linnets; and they are capable of propagation: nay, the breed ſeems to be improved by croſſing, for they are larger, ſtronger, and have clearer voices, &c. One effect of this intermixture is a multiplication of

* *Fringilla Carduelis*, var. 8. Linn.
Carduelis Hybrida, Briſſ.
The Canary Goldfinch, Alb. and Lath.

the

the pretended ſpecies. I ſhall give an example in the Siſkin.

The Hybrid of Albin was obtained from a cock Goldfinch, ſeparated in its tender age from the mother, and a hen Canary. It had the head, the back, and the wings, of the Goldfinch, but with a ſlighter tinge; the under-part of the body, and the quills of the tail, yellow, the laſt tipt with white. I have ſeen ſome whoſe head and neck were orange ; it ſeemed that the red of the male was melted into the yellow of the female.

The LULEAN FINCH*.

Le Chardonneret a Quatre Raies, Buff.
Fringilla Lulenſis, Linn.
Carduelis Suecica, Briſſ.

The moſt remarkable property of this bird is, that the wings, which are rufous-coloured at the baſe, are marked with four tranſverſe rays of different colours, and in this order, black, ruſt, black, white. The head, and all the upper-part of the body as far as the end of the tail, are of a dull cinereous ; the quills of the wings blackiſh ; the throat white ; the belly whitiſh, and the bill brown. This bird is found in the tract ſituated on the weſt of the gulph of Bothnia, near Lulhea.

* Specific character :—" Duſky ; the breaſt and ſhoulders rufous ; the wings black, with a rufous ſpot ; the under-ſurface of the neck and body white."

FOREIGN BIRDS, RELATED TO THE GOLDFINCHES.

I. The GREEN GOLDFINCH, or the MARACAXAO*.

Fringilla Melba, Linn.
Carduelis Viridis, Briss.

EDWARDS first figured and described this bird, which he tells us came from Brazil.—In the male the bill, the throat, and the interior part of the head are of a red more or less bright, except a small space between the bill and the eye, which is bluish; the hind-part of the head and neck, and the back, yellowish-green; the superior coverts of the wings, and their middle-quills, greenish, edged with red; the great quills almost black; the tail, and its superior coverts, of a bright red; the inferior coverts ash-gray; all the under-part of the body striped transversely with brown on a ground which is olive-green on

* Specific character: — "Green; the face and tail red; the "lower belly waved with white and black." Described thus by BRISSON:—"Above yellowish-green, below white, striated transversely with dusky; the anterior part of the head and the neck "scarlet colour; the breast olive-green; the tail-quills scarlet "above, and ash-coloured below."

on the breaſt, and continually grows fainter till it becomes entirely white under the belly. This bird is about the ſize of the Common Goldfinch; its bill is of the ſame ſhape, and its legs gray.

The female differs from the male, its bill being of a yellow fleſh-colour; the upper-part of its head and neck cinereous; the baſe of the wings and the rump yellowiſh-green, and ſo is the back, without any tint of red; the quills of the tail brown, edged exteriorly with red wine-colour; the inferior coverts white, and the legs fleſh-coloured.

M

II.

The YELLOW GOLDFINCH*.

Le Chardonneret Jaune, Buff.
Fringilla Triſtis, Linn.
Carduelis Americana, Briſſ. Klein.
American Goldfinch, Penn. Edw. Lath.

All thoſe who have mentioned this bird give it the appellation of American Goldfinch; yet this term would not be proper till it was aſcertained that no other Goldfinch exiſted in the

* Specific character: — "Yellow, forehead black, the wings "duſky." Thus deſcribed by Brisson: — "Yellowiſh Gold-"finch; crown black; white tranſverſe ſtripe on the wings; quills "of the wings and of the tail black; the exterior edges and the "tips of the leſſer wing-quills white." It feeds on thiſtle-ſeeds. The ſpecies includes a variety which is afterwards deſcribed by the name of the New York Siſkin.

 New

New World, and this ſuppoſition is not only improbable, but abſolutely falſe, for that of the preceding article is a native of Brazil; I have therefore adopted another denomination, which characterizes its plumage. The bill is nearly of the ſame ſhape and colour as the Common Goldfinch; the forehead black, which is peculiar to the male; the reſt of the head, the neck, the back, and the breaſt, ſhining yellow; the thighs, the lower belly, the ſuperior and inferior coverts of the tail, yellowiſh-white; the ſmall coverts of the wings yellow on the outſide, whitiſh on the inſide, and tipt with white; the great coverts black, and terminated with white ſlightly ſhaded with brown, which form two tranſverſe rays that are very apparent on the black wings; the middle quills of theſe are tipt with white; thoſe contiguous to the back and their coverts are edged with yellow; the quills of the tail amount to twelve, are of equal lengths, black above, and equal below; the lateral ones white on the inſide near the tip; the bill and legs fleſh-coloured.

In the female the forehead is not black, but of an olive-green, and all the upper-part of the body is of the ſame colour; the yellow of the rump, and of the under-part of the body, is leſs brilliant; the black on the wings more dilute, and on the contrary the tranſverſe rays are not ſo faint; laſtly, the belly, and the inferior coverts of the tail, are entirely white.

The

The young male is diſtinguiſhed from the female by nothing but its black forehead.

The female obſerved by Edwards was ſhut up alone in a cage, and yet layed in the month of Auguſt 1755, a ſmall egg of pearl-gray, and without ſpots; but, what is more uncommon, Edwards adds, that it moulted regularly twice a-year, in March and September. In winter the body was entirely brown; but the head, wings, and tail, retained that colour only in ſummer. The male died too ſoon for this obſervation; but probably, like the female, it would have dropt its feathers twice annually, and in that reſpect reſemble the Bengals, the Widow-birds, and many other natives of warm climates.

In the ſubject obſerved by Briſſon, the belly, the loins, the inferior coverts of the tail, and of the wings, were of the ſame yellow with the reſt of the body; the ſuperior coverts of the tail white-gray; the bill, the legs, and the nails, white: but moſt of theſe differences may be owing to the different ſtates in which the bird has been examined. Edwards drew it from the life, and his ſpecimen appears beſides to have been larger than that of Briſſon.

Cateſby tells us that it is very rare in Carolina, more frequent in Virginia, and very common in New York. The one figured in the *Pl. Enl.* was brought from Canada, where Father Charlevoix ſaw ſeveral of that ſpecies.

Total length four inches and one third; the bill five or ſix lines; the tarſus the ſame; the alar extent ſeven lines and one fourth; the tail eighteen lines, conſiſting of twelve equal quills, and ſtretching ſix lines beyond the wings.

The LESSER REDPOLL*.

Le Sizerin, Buff.
Fringilla Linaria, Linn, Brun. Kram. Frif. &c.
Fringilla Rubra Minor, Ray, Will. Briff. Klein.
The Leffer red-headed Linnet, or Redpoll, Penn. Barr. Ell. Alb. and Lath.

BRISSON calls this bird the Little Vine Linnet: but it appears to refemble the Sifkin, and its fong is much inferior to that of the Linnet. Gefner tells us, that at Nuremberg it has the name of *Tſchiit-Scherle*, on account of its fharp cry; he adds, that it appears only once in five or feven years †, like the Bohemian Chatterers, and arrives in immenfe flocks. We learn from the relations of voyagers, that it fometimes pufhes its excurfions as far as Greenland ‡. Frifch in-forms

* In German, *Zitzcherlein, Meer-Zeiſlein, Stock-henfling (Stick Linnet)*, and *Roth-plattige henfling, (Red Plated Linnet)*; in Swifs, *Schoperle*; in Pruffian, *Tchetzke*; in Swedifh, *Graoſiſka*; the Greek name is Αιγιθος.

† Whatever is uncommon turns into the marvellous. Some fay that the appearance of numerous flocks of the Leffer Redpolls forebodes a plague; others, that they are rats metamorphofed into birds before the winter, and refume their proper form in the fpring. In this way it is accounted for their not being found in fummer. SCHWENCKFELD.

‡ " Another bird appears in Greenland in the fummer, which " refembles the Linnet, though fmaller: it is diftinguifhed by its " head, which is partly red as blood; it may be fed with oat-meal

 " in

forms us, that in Germany it arrives in October and November, and departs in February.

I have ſaid that it bears more analogy to the Siſkin than to the Linnet; this was the opinion of Geſner, and it is alſo that of Dr. Lottinger, who is well acquainted with theſe little birds. Friſch goes farther, for he aſſerts that the Siſkin will ſerve for a call to allure the Leſſer Redpolls into ſnares in the time of paſſage, and that the two ſpecies intermix and propagate with each other. Aldrovandus perceived a ſtrong likeneſs between the Leſſer Redpoll and the Goldfinch, which, except its red head, reſembles much a Siſkin. A bird-catcher of great experience and little reading, told me that he has caught many of the Leſſer Redpolls intermixed with Siſkins, which they were very like, eſpecially the females, only their plumage was darker, and their bill ſhorter. Laſtly, Linnæus ſays, that the Leſſer Redpolls frequent places covered with alders, and Schwenckfeld reckons the ſeeds of theſe trees among the aliments which they prefer; but the Siſkins are extremely fond of theſe ſeeds. The Leſſer Redpolls

" in winter . . . Sometimes whole flocks of theſe birds alight on " board, like clouds driven by the wind, when a veſſel is eighty " or a hundred leagues from land. They have a pleaſant " ſong." *Continuation de l'Hiſt. des Voy.* May not thoſe be the ſame birds which the Chineſe breed in cages to fight " Theſe " birds reſemble Linnets, and as they perform diſtant journeys, " it will be the leſs ſurprizing to find them in a country ſo re- " mote!" NAVARETTE.

Redpolls eat not rape ſeed like the Linnet, but hemp ſeed, the ſeed of ſpeckled nettles, of thiſtles, of flax, of poppies, and crop the buds of young branches of oak, &c.: they mix readily with other birds: they are particularly tame in winter, and will then allow us to approach very near them without being ſcared *. In general, they have little timidity, and can eaſily be caught with lime-twigs.

The Leſſer Redpoll frequents the woods, and often lodges in the oaks: it creeps along the trunk like the Titmouſe, and alſo clings to the extremity of the ſmall branches. Hence probably is derived the name of *Linaria Truncalis* †, and perhaps that of Little Oak.

The Leſſer Redpolls grow very fat, and are excellent eating. Schwenckfeld ſays, that they have a craw like the poultry, diſtinct from the ſmall ſac formed by the dilatation of the *œſophagus* before its inſertion into the gizzard: this gizzard is muſcular, as is that of all the granivorous tribe, and many pebbles are found in it.

In the male, the breaſt and the top of the head are red, and there are two white tranſverſe ſtripes on the wings; the reſt of the head and all the upper-part of the body, mixed with brown and light rufous; the throat brown; the belly and the inferior coverts of the tail and

* Theſe remarks are Lottinger's. Schwenckfeld relates, that a prodigious number of the Leſſer Redpolls were caught in the beginning of winter A. D. 1602.

† i. e. Trunk Linnet.

wings,

wings, rufty white; their quills brown, with a complete border of a more delicate colour; the bill yellowifh, but brown near the tip; the legs brown. Thofe obferved by Schwenckfeld had cinereous backs.

In the female, there is no red except on the head, and it is befides lefs bright. Linnæus excludes it entirely; but perhaps the one which he examined had been kept long in the cage.

Klein relates, that having electrified in the fpring one of thefe birds, and a Goldfinch, without occafioning to them any fenfible injury, they both died the following October the fame night: but what deferves to be noticed is, that both had entirely loft their red tinge.

Total length above five inches; the alar extent eight inches and a half; the bill five or fix lines; the tail two inches and a half, and fomewhat forked, contains twelve quills, and projects more than an inch beyond the wings. [A]

[A] The fpecific character of the Leffer Redpoll (*Fringilla Linaria*, LINN.): – "Variegated with dufky and gray, above "tawny-white, whitifh double ftripe on the wings; the crown "and breaft red." Thus defcribed by Briffon: MALE. "Varie-"gated, above dufky and tawny gray, below tawny-white, dufky "fpots between the bill and the eyes and under the throat, the "crown and breaft red."—FEMALE. "Crown red, double "tranfverfe ftripe on the wings, tawny-white; the tail-quills "dufky, and the edges whitifh-gray."

It

It inhabits the whole extent of Europe, from Italy to the utmoſt verge of the Ruſſian empire: it is alſo found in the north of Aſia and America. It is only half the ſize of the Greater Redpoll. It builds its neſt among the alders, employing for that purpoſe ſmall ſticks and wool, and lining it with hairs and down. It lays four eggs of a light ſea-green colour, marked at the large end with reddiſh points.—It breeds in the north of England, and reſorts in flocks to the ſouthern counties in winter; and in that ſeaſon, it feeds principally on alder ſeeds.

In the female the ſpot on the head is ſaffron coloured, and not red.

The SISKIN*.

Le Tarin, Buff.
Fringilla Spinus, Linn. Gmel.
Acanthus Avicula, Gesner.
Ligurinus, Will. Briss.
Spinus, seu *Ligurinus*, Aldrov.
Siskin, or *Aberdavine*, Penn. and Lath.

Of all the granivorous birds, the Goldfinch is reckoned the most a-kin to the Siskin; both have the bill elongated and slender near the point; both are gentle, docile, and lively. The fruits of their intermixture are also fit to propagate.—Some naturalists have been induced by these analogies, to regard them as two contiguous species belonging to the same genus: indeed all the granivorous birds may be classed together; for their cross-breed are prolific. Since this general character extends to them all, it becomes the more necessary to select the distinguishing features, and to trace the precise boundaries of each species.

The Siskin is smaller than the Goldfinch; its bill is proportionally shorter, and its plumage is

* In German, *Zinsel*, *Zyschen*, *Zeislein*, *Engelchen*, *Zizing*, *Gruene Hensling* (Green Linnet); in Italian, *Lugaro*, *Lugarino*, *Luganello*, *Lucarino*; in Polish, *Czizeck*; in Turkish, *Utlugan*; in Swedish, *Siska*, *Groen Siska*; in Greek, Σπινος, Ακανθις, Θραυπις; in Latin, *Spinus*, *Acanthis*, *Thraupis*, and *Ligurinus*, from λιγυς on account of the shrillness of its notes.

entirely

entirely different: its head is not red, but black; its throat brown; the fore-part of its neck, its breaſt, and the lateral quills of its tail, yellow; the belly yellowiſh white; the under-part of the body olive-green, ſpeckled with black, which aſſumes a yellow caſt on the rump, and ſtill more yellow on the ſuperior coverts of the tail.

But in the more intimate qualities, which reſult directly from organization or inſtinct, the differences are ſtill greater. The Siſkin has a ſong peculiar to itſelf, and much inferior to that of the Goldfinch; it is very fond of alder-ſeeds, which the Goldfinch will never touch, and the Siſkin, in its turn, is indifferent about thiſtle-ſeeds: it creeps along the branches, and ſuſpends itſelf from their extremity like the Titmouſe:—In ſhort, we might regard it as an intermediate ſpecies to the Titmouſe and Goldfinch. Beſides, it is a bird of paſſage, and in its migrations it flies at a great height, and is heard before it can be ſeen; whereas the Goldfinch continues with us the whole year, and never flies very high: laſtly, theſe two birds are never obſerved to aſſociate together.

The Siſkin can be taught like the Goldfinch, to draw up the little bucket: it is equally docile, and though not ſo active, it is more cheerful; for it begins always the earlieſt in the morning to warble, and to rouze the other birds. But, as it has an unſuſpicious temper, it is eaſily decoyed into all ſorts of ſnares, traps, ſprings, &c.

and

and it is more eaſily trained than any other bird caught in the adult ſtate. We need only to offer it habitually the proper ſort of food in the hand, and it will ſoon become as tame as the moſt familiar Canary. We may even accuſtom it to perch upon the hand at the ſound of a bell; for if at firſt we ring at each meal, the ſubtle aſſociation of perceptions, which obtains alſo among the animals, will afterwards rouze it to the call. Though the Siſkin appears to ſelect its food with care, it conſumes much; but its voracious appetite is ſubordinate to a noble paſſion: it has always in the volery ſome favourite of its own ſpecies, or if that is not to be obtained, a bird of another ſpecies, which it cheriſhes and feeds with the fondneſs of a parent.— It drinks often *, but ſeldom bathes; it only approaches the margin of the water and dips its bill and breaſt, without much fluttering, except perhaps in hot weather.

It is ſaid that it breeds on the iſlands in the Rhine, in Franche-comté, in Switzerland, Greece, and Hungary, and that it prefers the mountain foreſts. Its neſt is very difficult to diſcover †, which

* The bird-catchers lay lime-twigs at the ſides of brooks, and are very ſucceſsful in the capture.

† " The bird-catchers in Orleans, ſays Salerne, agree that the " diſcovery of a Siſkin's neſt is a thing quite unheard of. It is " probable, however, that ſome continue in the country, and " breed near the banks of the Loiret, among the alders, of which " they are very fond; and the more ſo as young ones are ſome- " times caught with limed twigs or in traps. M. Colombeau " aſſures

which has given rise to a vulgar opinion, that the Siſkin renders it inviſible with a certain ſtone. Accordingly, our accounts are imperfect in regard to that ſubject: Friſch ſays that it conceals its neſt in holes; Kramer ſuppoſes the bird covers it with leaves, which is the reaſon that it is never found.—The beſt way to aſcertain the point, would be to obſerve how they proceed when they breed in a volery; which, though the trial has not hitherto ſucceeded, is ſtill poſſible.

But it is more common to croſs them with the Canaries. There ſeems to be a great ſympathy between the two ſpecies; they ſhew a reciprocal fondneſs at the very firſt meeting, and intermix indiſcriminately*. When a Siſkin is paired with a hen Canary, he eagerly ſhares her toils; he is buſy in carying materials for the neſt, and arranging them; and regularly diſgorges food for the ſitting female. But yet moſt of the eggs are addle: for the union of hearts is not alone ſufficient in generation, and the tem-

" aſſures me, that he found a neſt with five eggs in the bleachfield " of M. Hery de la-Salle." Kramer tells us, that in the foreſts ſkirted by the Danube, thouſands of young Siſkins are found, which have not dropt their firſt feathers, and yet it is very rare to meet with a neſt. One day when he was botanizing with one of his friends about the 15th of June, they both ſaw a male and female Siſkin often fly towards an alder with food in their bills; but, though they ſearched with all poſſible care, they could neither hear nor ſee the young ones.

* Father Bougot, from whom I received theſe remarks, has for five years ſeen a hen Canary breed thrice annually with the ſame cock Canary, and the four following years twice annually with another Canary, the firſt having died.

perament

perament of the Siſkin wants much of the warmth of the Canary.—The Hybrids reſemble both parents.

In Germany, the Siſkins begin to migrate in October, or even earlier; at this time they eat the hop ſeeds, to the great injury of the proprietors, and the places where they halt are ſtrewed with leaves. They entirely diſappear in December, and return in February *. In Burgundy, they arrive at the ſeaſon of vintage, and repaſs when the trees are in flower: they are particularly fond of the bloſſoms of the apple-tree.

In Provence, they leave the woods and deſcend from the mountains about the end of autumn. At that time, they appear in flocks of more than two hundred, and ſit all upon the ſame tree, or at a very little diſtance from each other. The paſſage continues fifteen or twenty days, after which ſcarcely any more are ſeen †.

The Siſkin of Provence is rather larger, and is of a finer yellow than that of Burgundy ‡.—It is a ſlight variety of climate.

Theſe birds are not ſo unfrequent in England as Turner ſuppoſed §. They are ſeen as in

* Friſch.

† Note of the Marquis de Piolenc.

‡ Note of M. Guys.

§ I mention this on the authority of Willughby. But the authors of the Britiſh Zoology ſay, that they never ſaw the bird in the country, and we muſt conclude that it is at leaſt rare in Britain.

other

other places, during their migration, and ſometimes they paſs in very numerous flocks, and at other times in very ſmall bodies. The immenſe flights happen only once in the courſe of three or four years, and ſome have ſuppoſed them to be brought by the wind *.

The ſong of the Siſkin is not diſagreeable, though much inferior to that of the Goldfinch, which it acquires, it is ſaid, with tolerable facility; it alſo copies the Canary, the Linnet, the Pettychaps, &c. if it has an opportunity of hearing them when young.

According to Olina, this bird lives ten years †; the female of Father Bougot has reached that age, but we muſt obſerve that in birds the females always outlive the males. However, the Siſkins are little ſubject to diſeaſes, except the melting of the fat, when they are fed with hemp ſeed.

The male Siſkin has the top of the head black, the reſt of the upper-part of the body olive, and ſlightly variegated with blackiſh; the ſmall upper coverts of the tail entirely yellow; the great coverts olive, terminated with cinereous; ſometimes the throat is brown, and even black ‡; the

* Olina. "In Pruſſia, myriads are caught in the yards." KLEIN.

† Thoſe which toil at the bucket (*à la galere*) are much ſhorter lived.

‡ All the adult males have not a black or brown throat: I have had ſome in which it was yellow, like the breaſt, and yet they had all the other characters of the males. I had an opportunity to

the cheeks, the fore-part of the neck, the breaſt, and the lower coverts of the tail, of a fine lemon yellow; the belly yellowiſh white; the flanks the ſame, but ſpeckled with black; there are two olive or yellow tranſverſe ſtripes on the wings, the quills of which are blackiſh, edged exteriorly with an olive-green; the quills of the tail yellow, except the two intermediate ones, which are blackiſh, edged with olive-green; they have all a black ſhaft; the bill has a brown point, the reſt white, and the legs are gray.

In the female, the upper-part of the head is not black, but ſomewhat variegated with gray; and the throat is neither yellow, brown, nor black, but white.

Total length, four inches and three-fourths; the bill five lines; the alar extent ſeven inches and two-thirds; the tail twenty-one lines, ſomewhat hooked, and projecting ſeven or eight lines beyond the wings. [A]

to ſee this black ſpot form by degrees on one caught in the net; it was at firſt about the ſize of a ſmall pea, and extended inſenſibly to a length of ſix lines, and a breadth of four, in the ſpace of eighteen months, and at preſent (8th April) it appears ſtill to grow. This Siſkin ſeems to be larger than common, and its breaſt of a finer yellow.

[A] Specific character of the Siſkin *Fringilla Spinus*:—"The "wing-quills are yellow in the middle, the firſt four ſpotleſs; the "quills of the tail yellow at the baſe, and black at the tip." The egg is very ſmall, and white, with reddiſh ſpots.

M

VARIETIES *of the* SPECIES *of* SISKINS.

I. In the month of September laſt year, a bird was brought to me that had been caught in a trap, and which muſt have been bred between the Siſkin and Canary; for it had the bill of the latter, and nearly the plumage of the former: it had undoubtedly eſcaped from ſome volery. I had no opportunity of hearing its ſong, or of obtaining progeny from it, ſince it died in March following; but M. Guys informs me, that in general the warble of theſe hybrids is varied and pleaſant. The upper-part of the body was mixed with gray, with brown, and with a little olive yellow; which laſt was the principal colour behind the neck, and was almoſt pure on the rump, on the fore-part of the neck, and of the breaſt as far as the thighs; laſtly, it bordered all the quills of the tail and wings, the ground of which was blackiſh, and almoſt all the ſuperior coverts of the wings.

Total length four inches and one-fourth; the bill three lines and a half; the alar extent ſeven inches and a half; the tail twenty-two lines, ſomewhat forked, and projecting nine lines beyond the wings; the hind-toe was the longeſt.

 The

The *æſophagus* two inches three lines, dilated in the ſhape of a ſmall pouch before its inſertion into the gizzard, which was muſcular, and lined with a looſe cartilaginous membrane; the inteſtinal tube ſeven inches and one-fourth; a ſmall gall-bladder, but no *cæcum*.

II. The NEW YORK SISKIN. We need only to compare this with the European Siſkin, to perceive that it is a variety reſulting from the difference of climate. It is rather larger, and has its bill ſomewhat ſhorter than ours; it has a black cap; the yellow of the throat and breaſt aſcends behind the neck, and forms a ſhort collar; the ſame colour borders moſt of the feathers on the higheſt part of the back, and appears again on the lower-part of the back and on the rump; the ſuperior coverts of the tail are white; the quills of the tail and of the wings are of a fine black, edged and tipt with white: all the under-part of the body is dirty white. As the Siſkins are roving birds, and fly very lofty, they may have migrated into North America, and ſuffered ſome changes in their plumage *.

III. The OLIVAREZ †. The upper-part of the body is olive; the under lemon; the head black;

* Mr. Latham reckons this bird a variety of the *Yellow Goldfinch*, or *American Goldfinch*, (Fringilla Triſtis,) before deſcribed.

† *Fringilla Spinus*, Var. 2. Linn.

black; the quills of the tail and wings blackiſh, edged more or leſs with light yellow; the wings marked with a yellow ſtripe. So far it much reſembles the European and the New York Siſkin, and its ſize and ſhape are the ſame. It is probably the ſame bird, which, being lately introduced into theſe different climates, has not yet undergone all the change.

In the female, the top of the head is of a brown-gray, and the cheeks lemon, as alſo the throat.

It has a pleaſant ſong, and in that reſpect excels all the birds of South America. It is found near Buenos-Ayres and the Straits of Magellan, in the woods which ſhelter it from the ſeverity of the cold and the violence of the winds. The one which Commerſon ſaw was caught by the foot between the two valves of a muſcle.

The bill and legs were cinereous; the pupil bluiſh; the mid-toe joined by its phalanx to the outer-toe; the hind-toe the thickeſt, and its nail the longeſt of all: it weighed an ounce.

Total length four inches and a half; the bill five lines; the alar extent eight inches; the tail twenty-two lines, ſomewhat forked, compoſed of twelve quills, and projecting about an inch beyond the wings; the wings conſiſt of only ſixteen feathers.

 IV. The

IV. The BLACK SISKIN *. As there are Black Goldfinches with an orange head, ſo there are Black Siſkins with a yellow head. Schwenckfeld ſaw one of that colour in the volery of a Sileſian gentleman; all the plumage was black except the top of the head, which was yellowiſh.

* *Fringilla Spinus*, Var. 2. Linn.
Ligurinus, Briſſ.

M

FOREIGN BIRDS, RELATED TO THE SISKIN.

I.

The CATOTOL*.

Fringilla Catotol, Gmel.
Cacatototl, Ray.
Ligurinus Mexicanus Niger, Briſſ.
The Black Mexican Siſkin, Lath.

THIS is the name given in Mexico to a ſmall bird of the ſize of our Siſkin, which has all the upper-part variegated with blackiſh fulvous, and all the lower-part whitiſh, and the legs cinereous: it reſides in plains, lives on the ſeeds of a tree called by the Mexicans *hoauhtli*, and ſings agreeably.

* Specific character:—"Variegated with blackiſh and fulvous, "below bright white."

 II. The

II.

The ACATECHILI*.

Fringilla Mexicana, Gmel.
Ligurinus Mexicanus, Briſſ.
Acatechichictli, Ray.
The Mexican Siſkin, Lath.

The little which we know of this bird evinces its relation to the Siſkin: its ſize is nearly the ſame; its ſong the ſame; and it feeds on the ſame ſubſtances: its head and all the upper-part of the body are greeniſh brown; the throat and all the under-part white ſhaded with yellow. The Mexican name *Acatechichictli*, ſignifies the *bird that rubs itſelf againſt the reeds*; may not this allude to ſome of its habits?

* Specific character:—"Greeniſh duſky, below whitiſh."

The TANAGRES.

Les Tangaras, Buff.

IN the warm parts of America is found a very numerous genus of birds, ſome of which are called *Tangaras* at Brazil; and nomenclators have adopted this name for all the ſpecies included. Theſe birds have been ſuppoſed by moſt travellers to be a kind of Sparrows; in fact, they differ from the European Sparrows only by their colours, and by a minute character, that the upper mandible is ſcolloped on both ſides near the point. They cloſely reſemble the Sparrows in their inſtinctive habits: they fly low and by jerks; their notes are for the moſt part harſh; they may be alſo reckoned granivorous, for they live upon very ſmall fruits; they are ſocial with each other, and, like the Sparrows, are ſo familiar as to viſit the dwellings: they ſettle in dry grounds, and never in marſhes; they lay two eggs, and ſometimes, though rarely, three.

The Sparrows of Cayenne have ſeldom more eggs, while thoſe of Europe have five or ſix; and this difference is perceived in general between birds of hot and thoſe of temperate climates. The ſmalneſs of the hatch is compenſated

ſated by its frequent repetition, love being cheriſhed and maintained by the continual and uniform warmth.

The whole genus of Tanagres, of which we know more than thirty ſpecies, excluſive of varieties, ſeems confined to the new continent; for all thoſe which we have received were brought from Guiana and other countries of America, and not from Africa or India. This multitude of ſpecies is not ſurpriſing; for, in general, the number of birds in the torrid zone is perhaps ten times greater than in other regions, becauſe nature is there more prolific, and leſs diſturbed in her operations by the interference of man; becauſe foreſts are there more frequent, ſubſiſtence is more plentiful, and the colds of winter are unknown: and the natives of the tropical countries, rioting in a perpetual abundance, are totally exempted from the riſks and dangers of a migration, and ſeldom are obliged even to ſhift from one haunt to another.

To avoid confuſion, we ſhall range the thirty ſpecies of Tanagres into three diviſions, adopting the characters from the moſt obvious difference, that of ſize.

The

N.º 98

THE GRAND TANAGRE

The GRAND TANAGRE.

Tanagra Magna, Gmel.

First Species.

This is represented N° 205, Pl. Enl. by the appellation of *the Tanagre of the Woods of Cayenne*; because I was told that it always came out of the extensive forests: but M. Sonini of Manoncour has since informed me that it also lodges often in the bushes in open situations. The male and female, which are much alike, commonly fly together. They live on small fruits, and sometimes eat the insects that prey on plants.

The figure will give a distinct idea of this bird. It is entirely a new species. [A]

[A] Specific character:—"Dusky olive; the forehead and "cheeks cœrulean; a black maxillary furrow; the neck and the "lower part of the rump red; the eye-brows, and a spot on the "throat white." It is of the size of a thrush; the under-part of the body reddish.

The CRESTED TANAGRE.

La Houppette, Buff.
Tanagra Cristata, Linn. and Gmel.
Tanagra Cayanensis Nigra Cristata, Briss.

Second Species.

This bird is not quite so large as the preceding, and is proportionally thicker. Its owes its name

name to a ſmall creſt which it can erect at pleaſure, and which diſtinguiſhes it from all the other Tanagres.

It is very common in Guiana, where it lives on ſmall fruits. It has a ſhrill cry, like that of the Chaffinch, but has not the ſong of that bird. It is found only in the cleared ſpots. [A]

[A] Specific character: — " Blackiſh; a gold-coloured creſt; " the throat and rump fulvous." Thus deſcribed by BRISSON: —" Creſted and blackiſh; the creſt gold-coloured; the feathers " at the baſe of the bill black; the throat, the loweſt part of the " back, and the rump, dilute fulvous; white ſpots on the wings; " the tail-quills blackiſh." It is ſix inches and one fourth long; its legs are lead-coloured.

The VIOLET TANAGRE.

Le Tangavio *, Buff.
Tanagra Bonarienſis, Gmel.

Third Species.

We are indebted to the late M. Commerſon for our knowledge of this bird: it is well preſerved in his collection: he had called it the *Black Bunting (Bruant Noir)*, which is very improper.—It is of a deep violet on the body, and even on the belly, with ſome greeniſh reflections on the wings and tail.

It meaſures from the end of the bill to that of the tail eight inches; its bill is blackiſh, and

* Contracted for *Tangara-violet*.

eight

eight or nine lines in length; its tail, which is not taper, is three inches long, and projects eighteen lines beyond the wings; the *tarsus* is about an inch long, and blackish, as well as the toes; the nails are thick and strong.

In the female the head is of a shining black, like polished steel; all the rest of the plumage is of an uniform blue. On the upper-part of the body, however, and on the rump, are some tints of a shining black.

The Violet Tanagre is found at Buenos-Ayres, and probably in other parts of Paraguay. We are unacquainted with its mode of life. [A]

[A] Specific character:—"Black-violet; the wings and tail "glossed with green."

The SCARLET TANAGRE.

Le Scarlatte, Buff.
Tanagra Rubra, var. Linn.

Fourth Species.

This bird is the same with the Cardinal of Brisson, and with the Scarlet Sparrow of Edwards. To it we should also refer, first, the two Red and Black Sparrows of Aldrovandus; the only difference being that the one happened to lose its tail, and this defect has been converted by Aldrovandus into a specific character, in which error he has been copied by all the ornithologists.

thologiſts*. Secondly, The *Tijepiranga* of Marcgrave†. Thirdly, The *Chiltototl*‡ of Fernandez. Fourthly, and laſtly, The Brazilian Blackbird of Belon, which received that name from thoſe who firſt imported it into France. Aldrovandus has copied Belon. The deſcriptions coincide in every reſpect, except in the ſongs of theſe birds; and I obſerved that thoſe which chanted were larger, had a brighter red tinge on the plumage, and alſo on the ſuperior coverts of the wings, &c. which makes it very probable that they were the males; indeed in almoſt all kinds of birds it is the males that are muſical.

It would alſo appear, that in the male the feathers on the head are longer, and form a ſort of creſt, as Edwards has figured it. This has led ſome travellers to ſay that there are two kinds of Cardinals in Mexico; one creſted, which ſings agreeably, and the other ſmaller, which ſings not at all.

Theſe birds belong to the warm climates of Mexico, Peru, and Brazil; but are rare in Guiana. Belon tells us that in his time the merchants who traded to Brazil drew conſiderable

* *Tanagra Braſilia*, var. 2. Gmel.
The Rumpleſs Blue and Red Indian Sparrow, Will.

† Mr. Latham reckons the *Tijepiranga* to be the female of the *Hooded Tanagre (Tanagra Pileata)*.

‡ This is the *Tanagra Braſilia* of Linnæus, the *Cardinalis* of Briſſon, and the *Braſilian Tanager* of Latham. Its ſpecific character:—"It is ſcarlet; its wings and tail black." It is ſix inches and one fourth long.

 profit

profit from the importation of them. Probably the feathers were employed to ornament the robes and other dreſſes then in faſhion, and theſe birds were more numerous than at preſent.

We may preſume that when travellers talk of the warble of the Cardinal, they mean the Scarlet Tanagre; for the *Creſted Cardinal* is of the genus of the Groſbeaks, and conſequently a ſilent bird. With regard to this point Salerne contradicts himſelf in the ſame page. It is univerſally admitted that this Tanagre has an agreeable warble, and is ſuſceptible even of inſtruction. Fernandez relates that it is found particularly at Totonocapa in Mexico, and ſings delightfully.

We reckon the following varieties of this ſpecies:—

Firſt, *The Spotted Cardinal* *, mentioned by Briſſon, which differs from the Scarlet Tanagre only becauſe ſome feathers of its back and breaſt are edged with green, which forms ſpots of the ſame colour, and of a creſcent ſhape. Aldrovandus calls this *The Short-tailed Blackbird.*

Secondly, *The Collared Cardinal* †, mentioned by Briſſon. It has not only the ſame ſize and colours as the Scarlet Tanagre, but the ſmall coverts, and the edges of the quills of the wings are blue, and on each ſide of the neck are two great ſpots of the ſame colour, they are conti-

* *Cardinalis Nævius*, Briſſ.

† *Cardinalis Torquatus*, Briſſ.

guous,

guous, and ſhaped like a creſcent. But Briſſon copies his deſcriptions of the Collared and Spotted Cardinal from Aldrovandus, who ſaw only the figures of theſe two birds, which renders their very exiſtence doubtful. Indeed I ſhould not have taken notice of them, did not the nomenclators inſert them in their catalogues.

Thirdly, The Mexican bird which Hernandez calls *The Parrot-coloured Mexican bird*, and which Briſſon deſcribes under the name of *Mexican Cardinal.* Hernandez ſays only, "This bird, from the lower part of the bill (which is ſomewhat hooked, and entirely cinereous) as far as the tail, including the whole of the belly, is of a minium tinge. The ſame colour is ſpread over the rump, and part of the back; but near the wings it receives a greeniſh tinge that gradually increaſes from thence to the neck, which is quite green. The head has an amethyſtine or hyacinthine tinge. The circle which ſurrounds the pupil is very white, and the orbits of a deep cœrulean. The origin of the wings is yellowiſh; their quills hyacinthine, and marked with a greeniſh ſtreak. The tail is entirely amethyſtine, without any mixture of green, and more dilute near the end. The legs, which have three toes before, and one behind, are of a cinereous-violet."

Theſe birds fly in flocks, and are eaſily caught with nooſes, and other ſnares. They are readily tamed; are fat, and good to eat.

The

The CANADA TANAGRE.

Tanagra Rubra, Gmel.
Cardinalis Canadensis, Briss.
The Red Tanagre, Penn. and Lath.
The Summer Red-bird, Catesby.

Fifth Species.

This bird differs from the Scarlet Tanagre by its size and plumage; it is smaller, and of a light flame-colour; its bill is entirely of a lead-colour, and has none of the peculiar characters; while in the Scarlet Tanagre, the upper-part of the bill is of a deep black, and the point of the lower mandible black, the rest of it white, and bellied transversely.

The Scarlet Tanagre is only found in the warmer parts of South America; as in Mexico, Peru, and Brazil. The Canada Tanagre occurs in many tracts in North America; in the country of the Illinois *, in Louisiana †, and in Florida ‡: so that there is no reason to doubt that these birds are of distinct species.

* "It is scarce more than an hundred leagues south of Canada that the Cardinals begin to be seen. Their song is sweet, their plumage beautiful, and their head wears a crest." CHARLEVOIX.

† Le Page Dupratz.

‡ "On Wednesday arrived at the port (of Havannah) a bark from Florida loaded with Cardinal-bird's skins and fruits . . . The Spaniards bought the Cardinal-birds at so high a price as ten dollars a-piece, and notwithstanding the public distress spent on them the sum of 18,000 dollars." GEMELLI CARERI.

It is accurately deſcribed by Briſſon. He has properly obſerved, that the red colour of its plumage is much lighter than in the Scarlet Tanagre. The ſuperior coverts of the wings, and the two quills next the body, are black; all the other quills of the wings are brown, and edged interiorly with white to their extremity; the tail conſiſts of twelve black quills, terminated by a ſmall border of light white; the lateral quills are rather longer than thoſe of the middle, which makes the tail ſomewhat forked. [A]

[A] Specific character of the *Tanagra Rubra*:—"It is red; "its wings and tail black; its tail-quills black at the tip." Thus deſcribed by BRISSON:—"Its wing-quills are duſky, their inner "edges white; the coverts of the wings and its tail-quills black, "the margin of the latter white at the tips."

The MISSISSIPPI TANAGRE.

Tanagra Miſſiſſippenſis, Gmel.

Sixth Species.

This is a new ſpecies. It reſembles much the Canada Tanagre, only its wings and tail are not black, but of the ſame colour with the reſt of the body. Its bill is larger and thicker than in any of the Tanagres; and alſo the mandibles are convex and inflated, which is uncommon even in any kind of the birds.—This character is badly expreſſed in the *Planches Enluminées*.

It

It is much inferior to the Scarlet Tanagre in point of ſong. It whiſtles ſo loud and ſo ſhrill that it would ſtun one in the houſe, and is fit only to be heard in the fields, or the woods. " In ſummer," ſays Dupratz, " we frequently hear this Cardinal in the foreſts, and in winter only on the banks of rivers after it has drank: during that ſeaſon it never quits its lodgment, but guards the proviſions which it has ſtored. Sometimes it collects as much as a Paris buſhel of maize, which it covers artfully with leaves, and then with ſmall branches or ſticks, and allows only a ſmall opening by which to enter into its magazine." [A]

[A] Specific character:—" It is entirely red."

The BLACK-FACED TANAGRE.

Le Camail, ou la Cravatte, Buff.
Tanagra Atra, Gmel.
Tanagra Melanopis, Lath.

Seventh Species.

This new ſpecies was preſented to the king's cabinet by Sonini de Manoncour. Its plumage is of an uniform cinereous; ſomewhat lighter under the belly, except the fore-part and the back of the head, of the throat, and of the top of the breaſt, which are ſpread with black. The wings and the tail are alſo cinereous, but deeper

 caſt

caſt than the upper-part of the body; the quills of the wings are edged exteriorly with a lighter aſh-colour, and thoſe of the tail with a ſtill more dilute ſhade.

This bird is the ſeventh of this genus in point of ſize. Its total length is ſeven inches; the bill nine lines; the upper mandible white at the baſe and black at the tip, the lower is entirely black; the tail is ſomewhat tapered, three inches and one fourth long, and projects two inches beyond the cloſed wings.

It is found in Guiana in the cleared ſpots, but is very rare, and has been noticed by no author. [A]

[A] Specific character: — "It is cinereous; the foreſide of its "head, and the whole of the lower part of its neck, are black."

The BLACK-HEADED TANAGRE.

Le Mordoré, Buff.
Tanagra Atricapilla, Gmel.

Eighth Species.

This is alſo a new ſpecies, and preſented by Sonini. It is of the ſame ſize with the preceding; its length ſeven inches; its head, wings, and tail, of a fine gloſſy black; the reſt of the body gilded dark brown, deeper on the fore-part of the neck and on the breaſt; its legs are brown; its tail, though tapered, is three inches long,

long, and projects fifteen lines beyond the wings; the bill is black, and nine lines long.

We are totally unacquainted with its habits. It is found in Guiana, and is ftill more rare than the preceding. [A]

[A] Specific character of the *Tanagra Atricapilla*:—"It is rufous-red; its head, wings, and tail, black; with a black furrow "on the wings."

The FURROW-CLAWED TANAGRE.

L'Onglet, Buff.
Tanagra Striata, Gmel.

Ninth Species.

The nails have on each fide a fmall furrow, running parallel to the edges. It was brought by Commerfon, and as it refembles the Tanagres in every other refpect, it is more than probable that it came from South America.

The head of this bird is ftriped with black and blue; the anterior part of the back is blackifh, and the pofterior bright orange; the upper coverts of the tail olive brown; the upper coverts of the wings, their quills, and thofe of the tail, are black, edged exteriorly with blue; all the under-part of the body is yellow.

Total length near feven inches; the bill eight lines, and furrowed near the point as in the Tanagres; the tarfus nine lines, and the mid-toe the fame.

 Com-

Commerſon has left no particulars with regard to its habits. [A]

[A] Specific character of the *Tanagra Striata*:—" It is black, " yellow below; its head ſtriped with cœrulean and black; the " loweſt part of its back orange."

The BLACK TANAGRE, and the RUFOUS TANAGRE.

Tenth Species.

Sonini informs us, that theſe conſtitute only one ſpecies, and that the one repreſented *Pl. Enl. No.* 179, *fig.* 2, is the male, and that of *No.* 711, the female. The female is entirely rufous, and the male entirely black, except a white ſpot on the top of each wing.—They are common in the cleared parts of Guiana; and, like the others, eat ſmall fruits, and ſometimes inſects. Their cry is ſhrill, and they have no ſong. They appear in pairs, and never in flocks.

The TURQUOISE TANAGRE.

Le Turquin, Buff.
Tanagra Braſilienſis, Linn.
Tanagra Braſilienſis Cærulea, Briſſ. and Klein.
Elototl, ſeu *Avis Spicæ Mayzii*, Ray.

Eleventh Species.

All the lower parts of the body, the upper-part of the head, and the ſides of the neck, are deep

deep or turquoiſe-blue; the forehead, the wings, and the tail, are black; there are alſo ſome ſpots of black near the legs, and a broad bar of the ſame below the breaſt.—This bird is found in Guiana, but is not frequent.

The RED-BREASTED TANAGRE.

Le Bec D'Argent, Buff.
Tanagra Jacapa, Linn.
Lanius Carbo, Pall.
Cardinalis Purpurea, Briſſ.

Twelfth Species.

The French ſettlers in Cayenne have given this bird the name of *Silver-Bill (Bec-d'Argent)*, which expreſſes a remarkable ſpecific character; viz. that the baſe of the lower mandible extends under the eyes, and forms on each ſide a thick plate, which, when the bird is alive, looks like the brighteſt ſilver; but this luſtre tarniſhes after death. It is imperfectly repreſented in the *Planches Enluminées*. Edwards has given an excellent figure of this bird under the name of *Red-breaſted Black-bird*; he is deceived indeed in regard to the genus, but he has hit the diſcriminating features.

The total length is ſix inches and a half, and that of the bill nine lines, which is black on the upper-part; the head, throat, and breaſt, are purple, and the reſt of the body black, with ſome purple tints. The iris is brown. The female differs from the male, not only in the colour of its bill, but in thoſe of its plumage; the upper-part of its body is brown with ſome ſhades of obſcure purple, and the under-part reddiſh; the tail and wings are brown.

Another diſcriminating character of the male, is a ſort of half collar round the occiput, formed by long purple briſtles, which project near three lines beyond the feathers. We are indebted to Sonini for this remark; and alſo for our acquaintance with this and all the other Tanagres of Guiana.

This bird is more numerous than any of the Tanagres in the Iſland of Cayenne and in Guiana; and it probably occurs in many other warm countries of America, for Fernandez gives the ſame account of a Mexican bird that frequents the vicinity of the mountains of Tepuzcullula. It feeds upon ſmall fruits, and alſo upon the large pulpy produce of the bananas, &c. when they are ripe; but eats no inſects. It haunts the cleareſt ſpots, and does not ſhun the neighbourhood of dwellings, and even viſits the gardens. However, the Red-breaſted Tanagres are alſo very common in deſert tracts, and even in the glades of the foreſts; for in ſpots where the trees are levelled by the hurricanes, and where the ſun darts his burning rays, there

are

are generally ſome of theſe birds, though always in pairs, and never in flocks.

Their neſt is cylindrical, and ſomewhat curved, which they faſten horizontally between the branches, the entrance being below; ſo that the rain, from whatever direction it beats, cannot penetrate. It is ſix inches long, and four inches and a half in diameter; it is conſtructed with ſtraws and the dry leaves of the Indian flowering reed*, and the bottom is well lined with broader portions of the ſame leaves: —it is generally fixed in the loftieſt trees. The female lays two elliptical eggs, which are white, and covered at the thick end with ſmall ſpots of light red, which melts away as it approaches the other end.

Some nomenclators have given this bird the name of Cardinal, but improperly: others have ſuppoſed that there is an obvious variety in this ſpecies. In Mauduit's cabinet we ſaw a bird whoſe plumage is pale roſe-colour, variegated with gray; I am rather inclined to think, that this difference is occaſioned by moulting. [A]

* *Canna Indica*, Linn.

[A] Specific character of the *Tanagra Jacapa*:—"It is black, "its front, throat, and breaſt ſcarlet." Thus deſcribed by Briſſon: *Male*, "dull purple; the quills of the wings and of the "tail, and the thighs, gloſſy black." *Female*, "above duſky, "mixed with dull purple; below tawny; the quills of the wings "and of the tail duſky." The Mexican name is *Chichiltototl*.

The

The SAINT DOMINGO TANAGRE.

L'Efclave, Buff.
Tanagra Dominica, Linn. Gmel. and Briff.

Thirteenth Species.

This Tanagre is called the *Slave* in Saint Domingo; and yet we are not told whether it can be bred in a cage, or is gentle and familiar as the name feems to import. Perhaps it owes the appellation to this circumftance;—the Crefted Fly-catcher in Saint Domingo, and the Forked-tail Fly-catcher of Canada, are termed *Tyrants*, and are much larger and ftronger than this bird, which alfo feeds on infects.

The Saint Domingo Tanagre bears fome refemblance to the Thrufhes; the colours, and particularly the fpeckles on the breaft, are fimilar in both, and, like the reft of its genus, it has the upper mandible fcalloped.

The head, the upper-part of the neck, the back, the rump, the fcapular feathers, and the fuperior coverts of the wings, are of an uniform colour; all the under-part of the body is of a dirty white, varied with brown fpots, that occupy the middle of each feather; the wing-quills are brown, edged exteriorly with olive, and interiorly with dirty-white; the two middle quills of the tail are brown, the reft of the

N.° 99

THE BISHOP TANAGRE.

the ſame colour, with an olive border on their inner-ſide; the tail is ſomewhat forked; the legs are brown. [A]

[A] Specific character of the *Tanagra Dominica*: — "It is "ſpotted with black, above duſky-olive, below whitiſh." Thus deſcribed by Briſſon: "Above duſky, below dirty-white, varie-"gated with duſky longitudinal ſpots; the quills of its wings and "of its tail duſky, their outer-edges olive."

The BISHOP TANAGRE.

Le Bluet, Buff.
Tanagra-Epiſcopus, Linn. Gmel. Briſſ. and Saler.
The Syacu, Edw.

Fourteenth Species.

It is larger than thoſe which form the ſecond diviſion of Tanagres*. In the male, all the upper-part of the body is bluiſh-gray; and in the female, all the upper-part of the head is of a yellowiſh-green, and all the upper part of the body, the back, the upper ſurface of the quills, the wings, and of the tail, olive-brown, gloſſed with violet; the broad bar on the wings, which is light-olive, is diſtinguiſhed from the brown on the back.

* Some ſentences are omitted here containing the author's reaſons for rejecting the appellation given this bird of *Cayenne Biſhop*, and for adopting that of *Bluet*.

Theſe

Theſe birds are very common in Cayenne; they haunt the ſkirts of the foreſts, plantations, and places that have been long cleared, where they feed upon ſmall fruits. They are never ſeen in large bodies, but always in pairs. They lodge at night among the leaves of the palm-trees, at their junction, near the ſtem, and make nearly the ſame noiſe that our Sparrows do among the willows; for they have no ſong, and their cry is ſharp and unpleaſant. [A]

[A] Specific character of the *Tanagra-Epiſcopus*:—It is cinere-"ous, its wings and tail cœrulean externally."

The RED-HEADED TANAGRE.

Le Rouge-Cap, Buff.
Tanagra Gularis, Linn. Gmel.
Cardinalis Americanus, Briſſ.

Fifteenth Species.

The head is tinged with a beautiful red; all the upper-part of the body is of a fine black; it has a narrow long ſpot of black on the breaſt, with purple ſpeckles; the legs and the upper mandible black; the lower mandible, yellow at the baſe and black at the tip.—The ſpecies is not very common in Guiana; nor are we certain whether it is found any where elſe. [A]

[A] Specific character of the *Tanagra Gularis*:—"It is black, "white below, its head red, its throat purple." Thus deſcribed by Briſſon: "Above gloſſy black, below ſnowy; the head and "the upper part of the throat ſcarlet; the tail-quills blackiſh."

The

The GREEN TANAGRE.

Le Tanagra Vert du Brefil, Buff.
Tanagra Virens, Linn. and Gmel.
Tanagra Brafilienfis Viridis, Briff.

Sixteenth Species.

This bird, which we know only from Briffon's defcription, is larger than the Houfe-Sparrow: all the upper-part of the body green; on each fide of the head is perceived a black fpot between the bill and the eye, under which is a bar of very brilliant beryl, that extends quite along the lower mandible; the fmalleft fuperior coverts are of a very brilliant fea-green, the others green.

The throat is of a fine black; the lower-part of the neck yellow, and all the reft of the under-part of the body yellowifh-green; the wings, when clofed, appear of a green running into blue; the quills of the tail the fame colour, except the two intermediate ones, which are green.

Briffon, to whom we are indebted for what we know of this bird, tells us, that it is found in Mexico, Peru, and Brazil. [A]

[A] Specific character of the *Tanagra Virens*:—" It is green, " yellow below, its ftraps and throat black, with a blue ftripe on " its jaw."

The OLIVE TANAGRE.

L'Olivet, Buff.
Tanagra Olivacea Gmel.

Seventeenth Species.

We have given this name, becaufe the plumage is of an olive-green, deeper on the upperpart of the body, and lighter on the under; the great quills of the wings have a ftill darker fhade, for they are almoft brown, and fhew only greenifh reflections.

Its length is almoft fix inches, and its wings reach to the middle of the tail.—It was brought from Cayenne by Sonini de Manoncour. [A]

[A] Specific character of the *Tanagra Olivacea* :—" It is olive, " its throat and breaft yellow, its belly white; the quills of its " wings and tail dufky, white at the edge." It is found alfo in New York.

The feventeen preceding fpecies form what we call *the Great Tanagres* :—we fhall now defcribe thofe which are of the medium fize, and which are not fo numerous.

The

The BLACK and BLUE TANAGRE.

Le Tanagra Diable-Enrhumé, Buff.
Tanagra Mexicana, Linn. and Gmel.
Tanagra Cayanensis Cœrulea, Briss.
The Black and Blue Titmouse, Edw.

First Middle Species.

The Creoles of Cayenne call this *The Rheum-Devil*: its plumage is mixed with blue, yellow, and black; the upper-part and sides of the head, the throat, the neck, and the rump, and the anterior part of the back, are black, without any tinge of blue; the small coverts of the wings are of a fine sea-green, and at the top of the wing take a violet cast; the last of these small coverts is black, terminated vith violet-blue; the quills of the wings black; the large ones (the first excepted) are edged exteriorly with green as far as the middle; the great coverts are black, edged exteriorly with violet-blue; the quills of the tail are black, slightly edged on the outside with blue-violet, as far as their ends; the first quill on each side has not this border, they are all gray below; a light yellow copper-colour is spread on the breast and belly, the sides of which and the coverts of the thighs are interspersed with black feathers tipt with violet-blue, and also with some yellow feathers spotted with black.

Total

Total length five inches and a half; the bill ſix lines; the tail an inch and ten lines, and ſtretches an inch beyond the wings.—It is found in Guiana, but is not frequent;—we are unacquainted with its hiſtory.

Briſſon thinks that this bird is the *Teoauhtototl* of Fernandez; but this naturaliſt only ſays, that it is about the ſize of a Sparrow, its bill ſhort, the upper-part of the body blue, and the under yellowiſh white, with black wings: from a deſcription ſo incomplete, it is impoſſible to decide the identity. Fernandez adds, that the *Teoauhtatotl* frequents the valleys and hills of *Tetzocan* in Mexico; that it is good eating; that its ſong is unpleaſant; and that it is not bred in houſes. [A]

[A] Specific character of the *Tanagra Mexicana*:—"It is "black, below yellowiſh, its breaſt and rump blue." Thus deſcribed by Briſſon: "Above gloſſy-black, below yellowiſh-white, "the ſides ſpotted with black and blue; the head, the lower-part "of the neck, the breaſt, and the rump blue; the quills of the "tail gloſſy-black."

The GRAY-HEADED TANAGRE.

Le Verderoux, Buff.
Tanagra Guianenſis, Gmel.

Second Middle Species.

The whole plumage of this bird is greeniſh, except the front, which is rufous from both ſides,

 on

on which two bars of the ſame colour extend from the front to the riſe of the red; the reſt of the head is aſh-gray.

Total length five inches and four lines; that of the bill ſeven lines, and that of the legs eight lines: the tail is not tapered, and the wings, when cloſed, do not quite reach the middle.

We are indebted to Sonini de Manoncour for this ſpecies, which is new. It is found in the extenſive foreſts of Guiana;—but we are unacquainted with its hiſtory.

The RUFOUS-HEADED TANAGRE*.

Le Paſſe-vert, Buff.
Tanagra Cayana, Linn. and Gmel.
Tanagra Cayanenſis Viridis, Briſſ.

The upper-part of the head is rufous; the upper-part of the neck, the lower-part of the back and the rump are of a pale-gold yellow, ſhining like raw ſilk, and in certain poſitions there appears a delicate tint of green; the ſides of the head are black; the higher-part of the back, the ſcapular feathers, the ſmall ſuperior coverts of the wings and thoſe of the tail are green.

The throat is blue-gray; the reſt of the under-part of the body ſhines with a confuſed

* This bird was by miſtake ranked among the Sparrows: it is now reſtored to its proper place.

mixture of pale-gold yellow, rufous, and blue-gray, and each of theſe predominates according to the light in which the bird is viewed; the quills of the wings and of the tail are brown, with a border of gold-green *.

In the female, the upper-part of the body is green, and the under of a dull-yellow, with ſome greeniſh reflexions.

Theſe birds are very common in Cayenne, where the Creoles call them *Dauphinois*; they inhabit only the cleared tracts, and even come near the plantations; they feed on fruits, and deſtroy vaſt quantities of bananas and Indian pears; they conſume alſo the crops of rice when in maturity; the male and female commonly follow each other, but they do not fly in flocks, only a number of them is ſometimes ſeen together among fields of rice.—They have no ſong or warble, and only a ſhort ſhrill cry. [A]

* In ſome individuals, the rufous at the top of the head deſcends much lower on the neck; in others, this colour extends on the one hand upon the breaſt and the belly, and on the other, upon the neck and all the upper-ſide of the body, and the green of the wing-feathers has a changing blue caſt.

[A] Specific character of the *Tanagra Cayana*: —" It is fulvous, its back green, its cap rufous, its cheeks black."

VARIETY.

VARIETY.

Le Passe-Vert à Tête Bleue, Buff.

Linnæus describes a bird resembling much the preceding. The fore-part of the neck, the breast, and the belly are golden-yellow; the back greenish-yellow; the wings and the tail green, without any mixture of yellow. It differs however in having its head of a bright blue.

The GREEN-HEADED TANAGRE.

Le Tricolor, Buff.
Tanagra Tricolor, Gmel.
Tanagra Cayanensis varia Chlorocephalos, Briss.

Fourth Middle Species.

Brought from Cayenne by Sonini. The plumage consists of three colours; red, green, and blue, which are all very bright. The two birds represented in N° 33 of the *Planches Enluminées*, seem to belong to the same species, and perhaps differ only in sex; for in the one the head is green and in the other blue; in the former, the upper-part of the neck is red, and in the latter green: — and these are almost the sole differences.

We have ſeen in the cabinet of M. Aubri, Rector of St. Louis, one of theſe in high preſervation, and ſaid to have come from the Straits of Magellan; but it is not very probable that the ſame bird ſhould inhabit the torrid climate of Cayenne, and the dreary frozen tracts of Patagonia.

The GRAY TANAGRE.

Le Gris-Olive, Buff.
Tanagra Griſea, Gmel.

Fifth Middle Species.

The under-part of the body is gray, the upper olive. It occurs both in Guiana and Louiſiana.

The PARADISE TANAGRE.

Le Septicolor, Buff.
Tanagra Tatao, Linn. and Gmel.
Avicula de Tatao, Seba.
Tanagra, Ray, Will. and Briſſ.
The Titmouſe of Paradiſe, Edw.

Sixth Middle Species.

The plumage is variegated with ſeven colours: fine green on the head, and the ſmall ſuperior coverts of the wings; gloſſy black on the upper-parts of the neck and back, on the middle quills of the wings, and on the upper ſur-

ſurface of the quills of the tail; brilliant fire-colour on the back; orange-yellow on the rump; violet-blue on the throat, the lower-part of the neck, and the great ſuperior coverts of the wings; deep gray on the under-ſurface of the tail; and, laſtly, fine ſea-green on all the under-part of the body from the breaſt. Theſe colours are all exceedingly bright, and well defined.

It does not aſſume the vivid red on the back till grown up, and the female never has that colour; the lower-part of her back too is orange like the rump, and in general her tints are more dilute, and not ſo diſtinctly defined as thoſe of the male —But there is ſtill ſome diverſity in the diſpoſition of the colours; ſome males have the bright red on the rump as well as on the back; and in many others both the back and rump are entirely of a gold colour.

The male and female are nearly of the ſame ſize, being five inches long; the bill only ſix lines, and the legs eight lines; the tail is ſomewhat forked, and the wings reach to the middle of it.

Theſe birds appear in numerous flocks. They feed upon the tender half-formed fruits which grow on a certain large tree in Guiana. They arrive in the iſland of Cayenne when this tree is in bloſſom, and depart ſoon afterwards, penetrating probably into the interior parts of the country when the ſame fruits are later in coming to maturity. They make their appearance in the inhabited parts of Guiana commonly about the middle of September, and ſtay about ſix weeks; they return again in April or May. Indeed they ſeem to ſeek always the ſame food; and when any of thoſe trees is in blow, we may certainly expect to find a number of theſe birds.

They breed not during their reſidence in Guiana. Marcgrave tells us that in Brazil they are kept in the cage, and fed on meal and bread. They have no warble, and their cry is ſhort and ſharp.

We muſt not with Briſſon range the *Talao* with this ſpecies; for the deſcription given by Seba is not at all applicable to it: "The Talao," ſays Seba, "has its plumage beautifully variegated with pale green, with black, with yellow, and with white; the feathers of the head and breaſt are finely ſhaded with pale green, and with black; and the bill, the legs, and the toes, are deep black." Beſides, what demonſtratively proves it to be not the ſame bird, the author adds, that it is very rare in Mexico; whereas the Paradiſe Tanagres we have ſeen arrive there in very great numbers.

The BLUE TANAGRE.

Tanagra Mexicana, var. Gmel.
Tanagra Barbadenſis Cærula, Briſſ.

Seventh Middle Species.

Its head, throat, and the under-part of the neck, are of a fine blue; the back of the head, the

No. 100

THE SMALL TANAGRE.

the upper-part of the neck, the back, the wings, and the tail, black; the ſuperior coverts of the wings black, and edged with blue; the breaſt, and the reſt of the under-part of the body, fine white.

On comparing this with what Seba calls the *American Sparrow*, they appear to be the ſame, differing only perhaps in age and ſex. Briſſon ſeems to have amplified the imperfect account of Seba; but as he does not produce his authorities, we cannot lay any weight on his deſcription.

Seba's bird came from Barbadoes; ours from Cayenne.

The BLACK-THROATED TANAGRE.

Eighth Middle Species.

This ſpecies is new. It was found in Guiana, and brought home by Sonini de Manoncour.

The head, and all the upper-part of the body, olive-green; the throat black; the breaſt orange; the ſides of the neck, and all the under-part of the body, fine yellow; the ſuperior coverts of the wings, the quills of the wings, and of the tail, brown, and edged with olive; the upper mandible black, the lower gray; and the legs blackiſh.

The HOODED TANAGRE.

La Coiffe Noire, Buff.
Tanagra Pileata, Gmel.

Ninth Middle Species.

The total length of this bird is four inches and ten lines; its bill is black, and nine lines long; all the under-part of the body is white, ſlightly varied with cinereous; the upper-part of the head is gloſſy black, which extends on each ſide of the neck in a black bar, diſtinctly marked on the white ground of the throat, which makes the bird look as if it were hooded with black. The quills of the tail are not tapered, and are all twenty-one lines long, and extend an inch beyond the wings; the legs are nine lines long.

The *Tiyepiranga* of Marcgrave, which Briſſon terms the *Cinereous Tanagre of Brazil*, would reſemble this bird exactly, if Marcgrave had mentioned the black hood: and this renders it probable that the one which we have deſcribed is the male, and that of Marcgrave the female of the ſame ſpecies.

They are found in Brazil and Guiana; but we are not acquainted with their hiſtory.

SMALL

SMALL TANAGRES.

The middle-ſized Tanagres which have been above enumerated, are in general not larger than a Linnet.—Thoſe which we are going to deſcribe are ſenſibly ſmaller, and exceed not the ſize of a Wren.

The RED-HEADED TANAGRE.

Le Rouverdin *, Buff.
Tanagra-Gyrola, Linn. and Gmel.
Tanagra Peruviana Viridis, Briſſ.
Fringilla Pectore Cæruleo, Klein.
Fringilla Viridis, Capite Spadiceo, Act. Petr.
The Red-headed Greenfinch, Edw.

Firſt Small Species.

Its head is green; its body entirely rufous, except a light blue ſpot on the breaſt, and a yellow ſpot on the top of the wing.

This ſpecies appears in many parts of South America; in Peru †, Surinam ‡, and Cayenne. It would ſeem that it migrates, for it is not found in the ſame place the whole year. It arrives in Guiana twice or thrice annually, to feed upon ſmall fruit that grows on a large tree, on which it perches in flocks; and again departs, probably after the proviſions are conſumed. As theſe birds are not frequent, and always avoid the

* Formed from *Roux-verd*. † Edwards. ‡ Briſſon.

the cleared and inhabited ſpots, their habits have not been obſerved. [A]

[A] Specific character of the *Tanagra-Gyrola*:—" It is green, " its head red, its collar yellow, its breaſt blue."

The SYACU TANAGRE.

Second Small Species.

The two birds repreſented in the *Planches Enluminées*, No. 133, fig. 1, No. 301, fig. 1, ſeem to belong to the ſame ſpecies, and differ perhaps only in the ſex. It is likely that the white-bellied one is the female, and the green-bellied one the male.

We give them the name of *Syacou*, contracted from the Brazilian appellation *Sayacou*; for we have no doubt that what Briſſon terms the *Variegated Tanagre of Brazil* is the ſame kind.

Theſe two birds wêre brought from Cayenne, where they are rare. [A]

[A] The laſt of theſe birds is the *Tanagra-Syaca* of Gmelin, the *Tanagra Braſilienſis Varia* of Briſſon. Its ſpecific character:—" Hoary, its wings ſomewhat blue." The firſt is the *Tanagra Punctata* of Gmelin, the *Tanagra Viridis Indica Punctata* of Briſſon, and the *Spotted Tanager* of Latham. Its ſpecific character:—" Green, dotted with black; below yellowiſh white."

The

The ORGANIST.

Third Small Species.

Such is the name this little bird receives at St. Domingo; becauſe it ſounds all the notes of the octave, riſing from the baſe to the treble. This ſort of ſong, which implies that the ear of this bird is organized ſimilarly to the human ear, is not only ſingular, but very pleaſant. The Chevalier Fabre Deſhayes has informed me in a letter, that in the ſouth of St. Domingo on the high mountains, there is a ſmall bird very rare and famous, called the *Muſician*, whoſe ſong can be written. We preſume that this is the ſame with the *Organiſt*. But ſtill we ſhould doubt of the regular ſucceſſion of muſical ſounds; for we had not the bird alive. It was preſented by the Count de Noë, who had brought it from the Spaniſh diſtrict of St. Domingo, where he told me it was very rare, and difficult to diſcover, or to ſhoot; becauſe it is ſhy, and artfully conceals itſelf; it even turns round the branch as the hunter changes place, to elude his view: ſo that though there be ſeveral of theſe birds on a tree, it often happens that not one of them can be perceived.

The length four inches; the plumage blue on the head and neck; the back, the wings, and the tail, are ſtained with black, running into

into coarſe blue; the forehead, the rump, and all the upper-part of the body, coloured with orange-colour.—This ſhort deſcription is ſufficient to diſcriminate it.

We find in Dupratz's Hiſtory of Louiſiana, the deſcription of a ſmall bird which he calls *Biſhop*, and which we believe to be the ſame with the *Organiſt*. "The Biſhop is a bird ſmaller than the Canary; its plumage is blue, verging on violet.—It feeds on many ſorts of ſmall ſeeds, among theſe *widlogouil* and *choupichoul*, a kind of millet peculiar to the country. Its notes are ſo flexible, its warble ſo tender, that when we once hear it, we become more reſerved in our eulogiums on this nightingale. Its ſong laſts during a *Miſerere*, and during the whole time it never makes an inſpiration; it reſts twice as long before it renews its muſic, the whole interval elapſed being about two hours."

Though Dupratz does not mention whether it gives the notes of the octaves as the Organiſt is ſaid to do, we cannot doubt their identity; for the colours and ſize are the ſame in both. The Scarlet Tanagre, which reſembles it in point of ſong, is twice as large; and the Arada, which has alſo a charming warble, is entirely brown. The Organiſt is then the only bird to which it can be referred.

The

The JACARINI TANAGRE.

Le Jacarini, Buff.
Tanagra Jacarina, Linn. and Gmel.
Tanagra Brasiliensis Nigra, Briss.
Carduelis Brasiliana, Will. and Edw.

Fourth Small Species.

This bird was called *Jacarini* by the Brazilians. Marcgrave mentions it, but takes no notice of its habits. However, Sonini de Manoncour, who observed it in Guiana, where it is very common, informs us that it prefers the cleared grounds, and is never seen in the large forests; that it lodges in the low trees, particularly the coffee-tree, and is distinguished by a singular circumstance, viz. that it springs from the branch on which it has perched a foot, or a foot and a half vertically, and falls back to the same spot; and thus continues to rise and sink alternately, till it removes to another bush, where it repeats the same exercise. Each leap is attended with a feeble cry, expressive of pleasure, and by an expansion of the tail. This would seem to be the mode in which the male courts the female; which on the contrary remains at ease, or hops about like other birds. The nest is composed of dry herbs of a gray colour; it is hemispherical, and two inches in diameter; the female deposits in it two elliptical eggs, seven or eight lines long, and of a greenish white, sprinkled with

with ſmall red ſpots, which are numerous, and ſpread moſt profuſely near the big end.

The Jacarini is eaſily known by its colour, which is black, and ſhining like poliſhed ſteel, and uniform over the whole body, except only in the male the interior coverts of the wings, which are whitiſh; for the female is entirely gray, and differs ſo much in plumage that it might be taken for a different ſpecies. The male alſo becomes gray in the moult. [A]

[A] Specific character of the *Tanagra Jacarina*:—"It is black-"violet, its wings whitiſh below, its tail wide-forked."

The GOLDEN TANAGRE.

Le Teité, Buff. Ray, and Will.
Tanagra Violacea, Linn. Gmel. Borowſk, &c.
Tanagra Braſilienſis Nigro-Lutea, Briſſ.
The Golden Titmouſe, Edw.

Fifth Small Species.

It is called *Teite* in its native region of Brazil. The female differs widely from the male; for the upper-part of the body is olive-green; the forehead, and the under-part of the bill, tinged partly with yellow, and partly with olive-yellow: whereas in the male the body is of a deep blue; and the forehead, the under-part of the throat and belly, fine yellow.

In the young bird the colours are ſomewhat different. The upper-part of the body is olive, ſprinkled with ſome feathers of a deep blue; and on the front the yellow is not diſtinctly marked. The feathers are only gray, with a little yellow at the tips; the under-ſide of the body is of as fine a yellow in the young bird as in the adult.

The ſame changes of plumage are obſerved in this as in the preceding ſpecies. The neſt is alſo very like that of the Jacarini, only it is not of ſo cloſe a texture, and is compoſed of reddiſh herbs inſtead of gray. There is a variety of it, which, as well as the ſpecies, is called *Little Louis* by the Creoles of Cayenne. They are both very common in Guiana, Surinam, and Brazil; they frequent the ground cleared near the farm-houſe, and feed on the ſmall fruits which they find on the buſhes; they light in great numbers among rice-crops, which muſt be guarded againſt their viſits.

They may be bred in the cage, where they are pleaſant, if five or ſix be put together. They whiſtle like the Bullfinch, and are fed on the plants called in Brazil *Paco* and *Mamao*. [A]

[A] Specific character of the *Tanagra Violacea*:—"It is violet; "below very yellow."

The

The NEGRO TANAGRE.

Le Tanagra Nègre, Buff.
Tanagra Cayanenſis, Gmel.
Tanagra Cayana, Linn.
Tanagra Cayanenſis Nigra, Briſſ.

Sixth Small Species.

This bird is of ſo deep a blue as to appear quite black, and it requires a cloſe inſpection to perceive ſome blue reflexions on its plumage; it has an orange-ſpot on each ſide of the breaſt, but covered by the wing; ſo that the general appearance is uniform black.

It is of the ſame ſize with the preceding, and inhabits the ſame countries, but is much rarer in Guiana. [A]

[A] Specific character of the *Tanagra Cayanenſis*:—"It is ſhining black; both ſides of its breaſt, and its wings, yellow."

Theſe are all the Great, the Middle-ſized, and the Small Tanagres, whoſe ſpecies can be aſcertained with accuracy. A few remain that have been deſcribed by Briſſon, but on the credit of authors whoſe accounts are vague and incorrect: I ſhall, however, enumerate them, without pretending to decide the ſpecies.

Firſt, *The Graſs bird*, or *Xiuhtototl* of Fernandez. All the body is blue, ſcattered with ſome fulvous feathers; thoſe of the tail black, and tipt with

with white; the under-part of the wings cinereous, the upper-part variegated with blue, with fulvous and with black; the bill short, somewhat thick, and of a rusty white; the legs are gray.

This author adds, that it is somewhat larger than our House-sparrow, that it is good eating, that it is raised in the cage, and that its song is not unpleasant.—It is impossible from such an imperfect account to decide whether it belongs to the genus of Tanagres. [A]

Secondly, *The Mexican Bird* of Seba, *of the size of a Sparrow*. The whole body is blue, varied with purple, except the wings, which are varied with red and black; the head is round; the eyes and the breast are covered above and below with a blackish down; the inferior coverts of the wings, and of the tail, are yellowish ash-colour. It is ranged among the singing birds.

So vague an account cannot warrant us to conclude that it belongs to the genus of the Tanagres; for the only points of analogy are that it inhabits Mexico, and is of the size of a Sparrow: and Seba's figure, as indeed all those of that author, can convey no distinct idea.

Thirdly, *The Brazilian Guira-Perea* of Marcgrave. It is about the bulk of a Lark; its bill black, short, and rather thick; all the upper-part of the body, and the belly, of a deep yel-

[A] This is the *Cærulean Tanagre* of Latham, the *Tanagra Canora* of Gmelin, and the *Tanagra Cærulea Novæ Hispaniæ* of Brisson.

 low,

low, ſpotted with black; the under-part of the head and neck, the throat, and the breaſt, black; the wings and tail compoſed of quills of blackiſh brown, and ſome edged exteriorly with green; the legs are of a dull cinereous.

It does not appear from this ſhort deſcription whether this bird ought to be referred to the Bulfinches or to the Tanagres. [A]

Fourthly, *The Bird ſmaller than the Goldfinch*, or *the Quatoztli of Brazil*, according to Seba. The half of its head is decorated with a white creſt; the neck is of a light red, and the breaſt of a fine purple; the wings deep red and purple; the back and the tail yellowiſh black, and the belly light yellow; the bill and legs are yellow. Seba adds, that it inhabits the mountains of *Tetzocano* in Brazil.

We ſhall obſerve, firſt, that the name *Quatoztli*, which Seba gives to this bird, is not Brazilian, but Mexican; and ſecondly, that the mountains of *Tetzocano* are in Mexico, and not in Brazil. It is probable therefore that he was miſtaken in calling it a Brazilian bird.

Laſtly, from the deſcription and the figure given by Seba, we ſhould rather range this bird in the genus of the Manakins than in that of the Tanagres. [B]

[A] This is the *Yellow Tanagre* of Latham, the *Tanagra Flava* of Gmelin, and the *Braſilienſis* of Briſſon.

[B] This is the *White-headed Tanagre*, or *Tanagra Albifrons* of Latham, the *Tanagra Leucocephala* of Gmelin, and the *Tanagra Braſilienſis Leucocephalus* of Briſſon.

 Fifthly,

Fifthly, *The Calatti* of Seba, which is nearly of the ſize of a Lark, and has a black creſt on the head; and the ſides of the head, and the breaſt, of a fine ſky-colour; the back is black, variegated with azure; the ſuperior coverts blue, with a purple ſpot; the quills of the wings variegated with green, with deep blue, and with black; the rump variegated with pale blue and green, and the belly with ſnowy white; the tail is of a beautiful form, brown terminated with rufous.

Seba adds, that this bird, which was ſent from Amboyna, is of an elegant figure (his plate is a very bad one), and that its ſong is alſo pleaſant. This is enough to exclude the *Calatti* from the Tanagres, which are found only in America, and in no part of the Eaſt Indies. [A]

Sixthly, *The Anonymous Bird* of Hernandez. The upper-part of its head is blue; the upper-part of the body is variegated with green and black, the under-part yellow, and ſpotted with white; the wings and the tail are deep green, with ſpots of lighter green; the legs are brown, and the toes and nails very long.

Hernandez ſubjoins in a corollary, that this bird has a black-hooked bill, and that if it were more curved, and if the toes were placed as in the parrots, he ſhould not heſitate to regard it as a real parrot.

[A] This is the *Amboina Tanagre* of Latham, the *Tanagra Amboinenſis* of Gmelin, and the *Tanagra Amboinenſis Cærulea* of Briſſon.

 From

From theſe indications, we ſhould refer this bird to the Shrikes.

Seventhly, *The Brown Cardinal* of Briſſon, which is a tropic bird, and not a Tanagre. [A]

[A] This is the *Military Tanagre* of Latham, the *Greater Bulfinch* or *Shirley* of Edwards, the *Tanagra Militaris* of Linnæus and Gmelin, and the *Cardinalis Fuſcus* of Briſſon.

The SILENT BIRD*.

We cannot refer this bird to any genus, and we place it after the Tanagres only becauſe its exterior appearance is ſimilar; but its habits are totally different. It never appears in the cleared ſpots, and remains always alone in the heart of the foreſts far from ſettlements, and has no ſong or cry. It rather hops than flies, and ſeldom reſts on the loweſt branches of the buſhes, for it commonly continues on the ground. It reſembles the Tanagres, however, in the ſhape of its body and feet, and in the ſlight ſcalloping on both ſides of the bill, which is longer than the bill of the Tanagres.—It is a native of the ſame climate of America.

* Latham ranges it with the Tanagres with the epithet *Silens*. Its ſpecific character:—" It is green; its head, and the under-" part of its body, hoary; its eye-brows, a fillet on its eyes, and " a ſtripe on its throat, black."

No. 101

THE BUNTING.

The ORTOLAN BUNTING*.

L'Ortolan, Buff.
Emberiza Hortulana, Linn. Gmel. Frif. Mull.
Hortulanus, Brif. Aldrov. &c.
Ortolano, Zinn. and Olin.

IT is very probable that our Ortolan is no other than the *Miliaria* of Varro, fo called becaufe it was fattened with millet feeds: it feems alfo to be the fame with the *Cenchramus* † of Ariftotle and Pliny, which is evidently derived from Κεγχρος, that likewife fignifies *millet*. And thefe etymological conjectures acquire force, from the correfpondence between the properties of thefe birds.

1. The *Cenchramus* is a bird of paffage, which, according to Ariftotle and Pliny, accompanies the Quails; in the fame way as the Rails, the Snipes, and other migratory birds ‡.

2. The *Cenchramus* utters cries during the night; which has given to thefe naturalifts occafion to fay, that it continually calls to the

* In German, *Fet-Ammer* (Fat-Bunting), *Jut-Vögel*; in Polifh, *Ogrodniczek*.

† Ariftotle fpells the name Κυγχραμος, not Κεγχαμος; Pliny indeed writes it both *Cynchramus* and *Cenchramus*: yet the etymology given in the text feems doubtful.

‡ Hift. Anim. *Lib*. viii. 12.—Hift. Nat. *Lib*. x. 23.

 companions

companions of its journey, and encourages them to advance.

3. Laſtly, in the time of Varro, the *Miliariæ*, as well as the Quails and Thruſhes, were fattened, and ſold at high prices to the luxurious *.

All these properties belong to our Ortolan:—It is a bird of paſſage, which a multitude of naturaliſts and of fowlers admit: it ſings during the night, as Kramer, Friſch, and Salerne affirm †: and laſtly, when fat, it is eſteemed a delicious morſel ‡. The Ortolans are not always caught fat; but there is an infallible method to bring them into that ſtate. They are ſhut up in a room from which the external light is excluded, but which is conſtantly illuminated with lanthorns, ſo that they cannot diſtinguiſh the night from the day: they are allowed to run about and pick up the oats and millet that are regularly ſcattered in the apartment. With this regimen they ſoon grow exceſſively fat, and if not prevented would even die of extreme corpulence §. When killed at the proper time, they are moſt delicate, delici-

* De Re Ruſtica, *Lib.* iii. 5.

† I could cite alſo the Sieur Burel, gardener at Lyons, who has ſometimes above an hundred Ortolans in his volery, and who communicated to me, or confirmed, many peculiarities of their hiſtory.

‡ It is pretended thoſe caught in the plains of Toulouſe are better flavoured than thoſe of Italy. In winter they are very ſcarce, and conſequently very dear: they are diſpatched to Paris by poſt in a ſmall trunk filled with millet, according to the hiſtorian of Languedoc, *tome* i. *p.* 46; in the ſame manner as they are ſent from Bologna and Florence to Rome in boxes full of corn, according to Aldrovandus.

§ It has been ſaid that they are ſometimes fattened to weigh three ounces.

ous

ous balls of fat, but rather too luſcious, and apt to cloy.—Thus Nature guards againſt intemperance by the ſatiety and diſguſt which conſtantly attend the exceſs of pleaſure.

The fat Ortolans are eaſily dreſſed in the water-bath, the ſand-bath, in hot aſhes, &c. And they may be alſo very well prepared in the ſhell of a real or an artificial egg, as was formerly practiſed with the beccafigos or epicurean warblers *.

It cannot be denied that the delicacy of their fleſh, or rather of their fat, has contributed more to their celebrity, than the charms of their warble. However, when kept in the cage, they have a ſong like that of the Yellow Bunting, which, as I have already obſerved, they repeat night and day. In countries where they are numerous, and conſequently well known, as in Lombardy, they are not only fattened for the table, but trained to ſing; and Salerne obſerves that there is a ſweetneſs in their notes. In this caſe they are better treated, and not being ſuffered to grow corpulent, their lives are prolonged. If they are kept a conſiderable time beſide other birds, they adopt ſomething of their ſong, eſpecially when they are young; but I know not whether they ever learn to articulate words, or catch the notes of our muſic.

Theſe birds make their appearance at the ſame time with our Swallow, or a little after, and they

* Petronius.

either accompany or precede the Quails. They enter Lower Provence, and advance as far as Burgundy, especially in the warm districts, which are planted with vineyards; however, they touch not the grapes, but eat the insects that prey upon the leaves and tendrils of the vines. On their arrival, they are rather lean, because then is the season of their loves *. They build in the vines, and their nests are pretty regular, and similar to those of the lark: they lay four or five grayish eggs, and have commonly two hatches in the year. In other countries, as in Lorraine, they place their nests on the ground, and prefer the corn-fields.

The young family begins to direct its course to the southern provinces as early as the first of August; but the parents do not commence their journey before the end of September. They remove into Fores, and halt in the neighbourhood of St. Chaumont and St. Etienne; they alight among the oats, of which they are very fond, and remain till the cold weather begins to set in; during which time they become so fat and inactive, that they might be knocked down with sticks. At this time they are proper for the table, especially the young ones; but they are more difficult to preserve than those which are caught in their first entrance. In Bearn also, the Ortolans appear twice annually; passing in May, and repassing in October.

* They may however be fatted at this time, by feeding them first with oats, then with hemp-seed, with millet, &c.

Some

Some have ſuppoſed theſe birds to be natives of Italy, whence they ſpread into Germany and other countries; and this is not improbable: for though they breed at preſent in Germany, they are caught promiſcuouſly with the Buntings and Chaffinches *; but Italy has been cultivated from a more remote antiquity. Beſides, it is not uncommon for theſe birds, when they diſcover a ſuitable country, to adopt it and ſettle in it. Not many years ſince, they were thus naturalized in a ſmall diſtrict of Lorraine, lying between Dieuſe and Mulée; where they breed, raiſe their young, depart in the fall, and return again in the ſpring †.

But their journies are not confined to Germany; Linnæus relates that they inhabit Sweden, and fixes the month of March for the period of their migration ‡. We muſt not however ſuppoſe that they are ſpread through all the countries between Sweden and Italy: they return conſtantly into our ſouthern provinces; ſometimes their courſe lies through Picardy, but ſcarcely are they ever ſeen in the northern part of Burgundy where I live, in Brie, or in Switzerland, &c. § They may be caught either with the nooſe, or with limed-twigs.

* Friſch.—Kramer ranks them among the birds which occur in Lower Auſtria; and he adds, that they live in the fields, and perch upon the trees which grow in the midſt of meadows.

† Dr. Lottinger. ‡ Fauna Suecica. § Geſner.

In

In the male, the throat is yellowifh, edged with cinereous; the orbits alfo yellowifh; the breaft, the belly, and the fides rufous, with fome fpeckles, whence the Italian name *Tordino* *; the lower coverts of the tail of the fame colour, but lighter; the upper-part of the body variegated with brown-chefnut, and blackifh; the rump and the fuperior part of the coverts of the tail, uniform brown-chefnut; the quills of the wings blackifh, the large ones edged exteriorly with gray, the middle ones with rufous; their fuperior coverts variegated with brown and rufous; the inferior ones with fulphur-yellow; the quills of the tail blackifh, edged with ruft, the two outer ones edged with white; laftly, the bill and legs are yellowifh.

The female has rather more cinereous on the head and on the neck, and no yellow fpot below the eye; but, in general, the plumage of the Ortolan is fubject to many varieties.

The Ortolan is fmaller than the Houfe Sparrow. Length, from fix inches and one-fourth, to five inches and two thirds; the bill five lines; the leg nine lines; the middle toe eight lines; the alar extent nine inches; the tail two inches and a half, confifting of twelve quills, and projecting eighteen or twenty lines beyond the wings. [A]

* From *Tordo*, a Thrufh.

[A] Specific character of the Ortolan, *Emberiza-Hortulana*:— "Its wing-quills dufky, the firft three whitifh at the margin; the

" tail-

" tail-quills duſky, the two lateral ones black exteriorly." It is ſix inches and one-fourth long. It builds in low buſhes, or on the ground, a neſt like that of the Sky-lark, and lays four or five gray eggs.

VARIETIES of the ORTOLAN.

I. The YELLOW ORTOLAN. Aldrovandus, who obſerved this variety, ſays that its plumage was ſtraw-coloured, except the wing-quills, which were tipt with white, and the outermoſt edged with the ſame colour: another peculiarity, the bill and legs were red.

II. The WHITE ORTOLAN. Aldrovandus compares its whiteneſs to that of the ſwan, and ſays, that all its plumage was uniformly the ſame. Burel of Lyons, who has long been employed in raiſing Ortolans, aſſures me that he has ſeen ſome grow white from age.

III. The BLACKISH ORTOLAN. Burel has ſeen others, which were undoubtedly of a different character from the preceding, and which became blackiſh as they grew old. In the one obſerved by Aldrovandus, the head and neck were green with a little white on the head and on two quills of the wing; the bill was red, and the legs cinereous; all the reſt was blackiſh.

IV. The

IV. The White-tailed Ortolan. It differs from the common Ortolan by the colour of its tail, and by the tints of its plumage, which are fainter.

V. I have obferved one, in which the throat was yellow, mixed with gray; the breaft gray, and the belly rufous.

The REED BUNTING*.

L'Ortolan de Roſeaux, Buff.
Emberiza Schœniclus, Lin. Gmel. Brun. Kram. &c.
Paſſer Torquatus, ſeu *Arundinaceus*, Ray, Will. Briſſ.
Emmerling, Gunth. Neſt.

UPON comparing the different birds of this family, I have perceived ſo many ſtriking relations between that of the preſent article and thoſe of the four following, that I ſhould have referred them all to the ſame ſpecies, if I could have collected a ſufficient number of facts to juſtify this innovation. It is extremely probable that all theſe would propagate with each other, and that the croſs-breed would be prolific. But ſtill I obſerve that they continue for a length of time in the ſame country without intermixing; that they retain their diſcriminating characters; and that their inſtincts are not quite the ſame. I ſhall therefore follow the common diviſion, but proteſt againſt the multiplication of the number of ſpecies, which is ſo fertile a ſource of error and confuſion.

* In Greek, Σχοινικλος, Σχοινικος, Σχοινιων (from Σχοινος, *a ruſh*); Στρουθος Σχοινικλος, (*Ruſh-Sparrow*); Βαλις (perhaps from Βαλος, a *thorn*); in German, *Rhor-Spar*, *Rhor-Sperling*, or *Rhor-Spatz*, (i. e. Reed Sparrow); *Weiden-Spatz* (Willow Sparrow); in Swediſh, *Saefsparf*; in Poliſh, *Wrobel Trzcinnis*.

The

The Reed Ortolans delight in fens, and neſtle among the ruſhes; however, they ſometimes reſort to the high grounds in rainy ſeaſons. In ſpring they are ſeen by the ſides of the roads, and in Auguſt they feed in the corn fields. Kramer tells us that they are fondeſt of millet. In general they ſeek their food like the Buntings, along the hedges, and in the cultivated ſpots. They keep near the ground, and ſeldom perch except on the buſhes. They never aſſemble in flocks, and ſcarcely more than three or four are ſeen at once. They arrive in Lorraine about the month of April, and depart in autúmn; however, there are always ſome of them which continue in that province during the winter.—They are found in Sweden, Germany, England, France, and ſometimes in Italy, &c.

This little bird is almoſt perpetually on the watch, to diſcover its foe, and when it deſcries ſome fowlers, it makes an inceſſant cry, which is not only teazing, but ſometimes ſcares away the game. I have ſeen ſportſmen quite out of patience at the notes of this bird, which are ſomewhat like thoſe of the Sparrow. The Reed Bunting has beſides a pleaſant warble in the time of hatching, about the month of May.

This bird jerks its tail upwards and downwards as quick as the Wagtails, and with more animation.

In

In the male, the upper-part of the head is black; the throat and the fore-part of the neck variegated with black and rusty gray; a white collar on the upper-part of the neck only; a sort of eye-brow and a bar of the same colour under the eyes; the upper-part of the body variegated with rufous and black; the rump and the superior coverts of the tail variegated with gray and rusty; the under-part of the body white, shaded with rusty; the sides slightly spotted with blackish; the quills of the wings brown, edged with different shades of rufous; the quills of the tail the same, except the two outermost on each side, which are edged with white; the bill is brown, and the legs are of a dark flesh-colour.

The female has no collar; its throat is not so black, and its head is variegated with black and light rufous: the white which occurs in its plumage is not pure, but always sullied with a rufous cast.

Length, from five inches three-fourths to five inches; the bill four lines and a half; the leg nine lines; the middle toe eight lines; the alar extent nine inches; the tail two inches and a half, consisting of twelve quills, and projecting about fifteen lines beyond the wings. [A]

[A] Specific character of the Reed Bunting, *Emberiza Schœniclus*:—"Its head is black, its body gray and black, the outer-"most quills of its tail marked with a wedge-shaped white spot." It is five inches and three-fourths long. It is found as high as Denmark,

Denmark; it is frequent in the ſouthern parts of Ruſſia; and it viſits Britain in the ſummer. It ſuſpends its neſt between four reeds, a few feet above the water; this neſt is made of graſs-ſtalks, lined with the down of reeds. The bird lays four or five eggs of a bluiſh white, variegated with purple ſtreaks. It ſings, eſpecially at night.

M

The HOODED BUNTING*.

Coqueluche, Buff.

A ſort of hood of a fine black covers the head, throat, and neck, and then, tapering to a point, deſcends on the breaſt, nearly in the ſame manner as in the Reed Ortolan; and this black is never interrupted, except by a ſmall white ſpot on each ſide, very near the opening of the bill; the reſt of the under-part of the body is whitiſh, but the ſides are ſpeckled with black. The hood which I have mentioned is edged with white behind; all the reſt of the upper-part of the body variegated with rufous, and blackiſh; the quills of the tail are blackiſh, but the two intermediate ones are edged with ruſty; the two outermoſt have a large oblique ſpot; the three others are uniform throughout.

Total length five inches; the bill ſix lines, entirely black, the tarſus nine lines; the tail two inches, rather forked, and projects about thirteen lines beyond the wings.

* This bird is in the cabinet of Dr. Mauduit, who has called it the Siberian Reed Ortolan. I have not ventured to adopt this appellation, leſt the ſubject ſhould be found to be only a variety of our Reed Ortolan.

M

The

The MUSTACHOE BUNTING.

Le Gavoué de Provence, Buff.
Emberiza Provincialis, Gmel.

This bird is diſtinguiſhed by a black ſpot at the ears, and a line of the ſame colour which extends on each ſide of the bill like muſtachoes. The lower-part of the body is cinereous; the upper-part of the head and of the body, variegated with rufous and blackiſh; the quills of the tail are divided by the ſame colours, the rufous being exterior and apparent, and the blackiſh being within and concealed. There is alſo a little whitiſh round the eyes, and on the great coverts of the wings.

This bird feeds on grain; it is fond of perching; and in the month of April its ſong is pleaſant.

This is a new ſpecies introduced by M. Guys.

Total length four inches and three-fourths; the bill five lines, the tail twenty lines, ſomewhat forked, and ſtretching thirteen lines beyond the wings. [A]

[A] Specific character of the *Emberiza Provincialis*:—"It is "variegated with gray and black; the under-ſide of the body "and a ſpot on its wings, white; a ſpot under the eye, and ſtripe "on the jaw, black."

M

The LESBIAN BUNTING.

Le Mitilene de Provence, Buff.
Emeriza Lesbia, Gmel.

This bird differs in ſeveral reſpects from the preceding: the black which appears on the ſides of the head is diſpoſed in three narrow ſtripes, parted by white ſpaces; the rump and the ſuperior coverts of the tail are ſhaded with ſeveral rufous tints. But the difference of habits affords a more marked diſtinction; the Leſbian Bunting does not begin its ſong before the month of June; it is more rare and ſhyer, and its repeated cries warn the other birds of the approach of the Kite, of the Buzzard, or of the Hawk; in which circumſtance it reſembles the Reed Bunting. The preſent inhabitants of Mytilene, or ancient Leſbos, employ it, on this account, as a ſentinel for their poultry, but take the precaution to ſhut it in a ſtrong cage, leſt itſelf become the prey of the ferocious invader. [A]

[A] Specific character of the *Emberiza Leſbia*:—" It is variegated with gray and black; its under-ſurface and its orbits white; has three ſtripes of black and white under its eyes; its lateral tail-quills partly white."

M

The

The LORRAINE BUNTING.

L'Ortolan de Lorraine, Buff.
Emberiza Lotharingica, Gmel.

Lottinger ſent us this bird from Lorraine, where it is very common.—Its throat, the fore-part of its neck, and its breaſt, are of a light aſh-colour, ſpeckled with black: the reſt of the under-part of its head and body rufous, ſpeckled with black; the ſpace round the eyes of a lighter colour; there is a black ſtreak below the eyes: the ſmall coverts of the wings are of light cinereous without ſpeckles; the others parted by rufous and black; the firſt quills of the wings black, edged with light cinereous, the following with rufous; the two middle quills of the tail rufous, edged with gray, the others partly black and partly white, but the outermoſt have always a greater ſhare of white; the bill is rufous-brown, and the legs of a lighter ſhade.

Total length ſix inches and a half; the bill five lines and a half; the tail two inches and four lines, and exceeds the wings by fifteen lines.

The female has a ſort of collar mottled with rufous and white; all the reſt of the under-part of the body is ruſty-white; the upper-part of the head is variegated with black, with rufous, and with white, but the black diſappears behind the head, and the rufous grows more dilute, ſo

 that

that an almoſt uniform ruſty-gray is produced. It has white eye-brows; the cheeks are deep rufous; the bill orange-yellow at its baſe, and black at its point; the edges of the lower mandible are adapted into the upper; the tongue is forked, and the legs are black.

One of theſe birds was brought to me the 10th of January; it had been juſt killed on a ſtone in the middle of the high-road; it weighed an ounce; its inteſtines meaſured ten inches; it had two very ſmall *cæca*; the gizzard was very thick, about an inch long, and ſeven lines and a half broad, and filled with vegetable ſubſtances, and many ſmall pebbles; the cartilaginous membrane in which it was ſheathed, had more adheſion than is uſual in birds.

Total length five inches ten lines; the bill five lines and a half; the alar extent twelve inches; the tail two inches and a half, ſomewhat forked, and projecting about an inch beyond the wings; the hind nail four lines and a half longer than the toe.

The LOUISIANE BUNTING.

L'Ortolan de la *Louifiane,* Buff.
Emberiza Ludovicia, Linn. Gmel.
Hortulanus Ludovicianus, Briff.

This American bird has the fame mottling of whitifh and black on its head that is common to almoft all our Ortolans; but its tail, inftead of being forked, is on the contrary fomewhat tapered. The top of the head is marked with a black horfe-fhoe, which opens befide the bill, and its branches extending above the eyes, coalefce at the back of the head: there are fome other irregular fpots below the eyes; rufous predominates on all the lower-part of the body, being deeper on the breaft and lighter above and below it; the upper-part of the body is variegated with rufous and black, and fo are the great and middle coverts and the quills of the wings next the body; but all the other quills and the fmall coverts of the wings are black, as well as the rump, the tail, and its fuperior coverts; the bill has blackifh fpots on a rufous ground; the legs are cinereous.

Total length five inches and one-fourth; the bill five lines; the tail two inches and one-fourth, confifting of twelve quills, fomewhat taper, and projects fourteen lines beyond the wings. [A]

[A] Specific character of the *Emberiza Ludoviciana*:—" It is " rufous fpotted with black, below pale, the breaft rufous, the head " marked with a black arch."

M

The YELLOW-BELLIED CAPE BUNTING.

L'Ortolan à Ventre Jaune du Cape de Bonne-Esperance, Buff.
Emberiza Capensis, var. B. Linn.

We received this from Sonnerat. It is the most beautiful of the genus; its head is of a glossy black, with five white rays nearly parallel, the middle extending to the lower-part of the neck; all the upper-part of the body is yellow, but deepest on the breast, from which it spreads upwards and downwards, growing more dilute by imperceptible degrees, so that the origin of the neck, and the last of the inferior coverts of the tail, are almost white; a gray transverse bar separates the neck from the back, which is brown-rufous, variegated with a lighter colour; the rump is gray; the tail brown, edged with white on both sides, and delicately tipt with the same; the small coverts of the wings ash-gray; the uncovered part of the middle ones white; the great ones brown, edged with rufous; the quills of the wings blackish, edged with white, except those next the body, which are edged with rufous; the third and fourth are the longest of all. With respect to the quills of the tail, the outermost, and the one next it on each side, are shorter than the rest; so that were the tail equally divided, though the whole be somewhat forked, each of the parts is taper; the greatest difference between the length of the quills is three lines.

In

In the female the colours are leſs vivid and diſtinct. Total length ſix inches and one fourth; the bill ſix lines; the tail two inches and three-fourths, conſiſting of twelve quills, and exceeding the wings fifteen lines; the *tarſus* eight or nine lines; the hind nail is the ſtrongeſt of all.

M

The CAPE BUNTING.

L'Ortolan du Cap de Bonne Eſperance, Buff.
Emberiza Capenſis, Linn. and Gmel.
Hortulanus Capitis Bonæ Spei, Briſſ.

This bird is as remarkable for duſky ill-defined colours, as the preceding is conſpicuous for the richneſs and luſtre of plumage: it has, however, two black ſtreaks, the one above, and the other below the eyes, which characterize the genus. The upper-part of its head and neck is variegated with dirty gray, and blackiſh; the upper-part of the body black, and yellowiſh-rufous; the throat, the breaſt, and all the under-part of the body, dirty gray; the ſuperior coverts of the wings rufous; the great coverts, and the quills of the wings, and the quills of the tail, blackiſh, edged with ruſty; the bill and legs blackiſh.

Total length five inches and three-fourths; the bill five lines; the alar extent near nine inches; the tail two inches and a half, conſiſting of twelve quills, and exceeds the wings by fifteen lines. [A]

[A] Specific character of the *Emberiza Capenſis*.—"It is gray, "its throat whitiſh, a ſtripe on its eyes, and cheeks blackiſh."

M.

The SNOW BUNTING*.

L'Ortolan de Neige, Buff.
Emberiza Nivalis, Linn. Gmel. Scop. Mull. Frisch. &c.
Emberiza Varia, Klein.
Hortulanus Nivalis, Briss.
Avis Nivis, Mart. Spitz.
See-Lerche, Wirs.
Pied Mountain Finch, Alb.
Sea-Lark, Ray.

THE mountains of Spitzbergen, the Lapland Alps, the shores of Hudson's-bay, and perhaps countries still more northerly, are, during the summer months, the favourite abodes of this bird. The excessive severity of these inhospitable climates changes part of its plumage into white in winter †. It has some variety of appearance from the diversified intermixture of white, black, or rusty; and the combinations of these colours are affected by the season, and by the temperature of the air.

* In Polish, *Sniegula, Snieznniczka*; in Danish, *Sneekok, Winter-fugel*; in Swedish, *Snoesparf*; in Norwegian, *Snee-fugl, Fiælster, Snee-spurre, Snee-titing, Sælskriger*; in Icelandic, *Sino-tytlingur, Soel-skrikia, Tytlings-blike*; in Laplandic, *Alpe, Alaipg*; in Dalecarlien, *Illwars-vogel*; in Scanian, *Sioe laerka*; in Greenlandish, *Kop-ano-arsuch*.

† Those white feathers are black at the base; and sometimes the black shines through the white, and forms a multitude of little spots, as in the individual painted by Frisch, under the name of *The Spotted White Bunting*. At other times the black collar of the base of each feather extends on a great part of the wing; so that a blackish colour is thus produced over all the under-side of the body, as in the Blackish and Yellowish Finch of Aldrovandus. *Lib. XVIII. pp.* 817, 818.

In

In winter the head, the neck, the coverts of the wings, and all the under-part of the body, are in the male white as ſnow, with a light and almoſt tranſparent tint of ruſty on the head only; the back black; the quills of the wings, and of the tail, partly black, partly white. In ſummer the head, the neck, the under-part of the body, and even the back, are ſtained with tranſverſe ruſty waves of various intenſity, but never ſo deep as in the female, of which it is the predominant colour, and diſpoſed in longitudinal ſtripes. In ſome ſubjects the neck is cinereous, the back cinereous variegated with brown; a purple tinge round the eyes; a reddiſh caſt on the head*, &c. The colour of the bill is alſo variable; ſometimes yellow, ſometimes cinereous at the baſe, and generally black at the point. In all, the noſtrils are round, ſomewhat prominent, and covered with ſmall feathers; the tongue is a little forked; the eyes are ſmall and black; the legs black, or blackiſh.

Theſe birds leave their mountains when the ſnows and froſts will not permit them to procure their food. This is the ſame with that of the white grous, and conſiſts of the ſeeds of a ſpecies of birch†, and of other ſeeds. When kept in the cage they are very well reconciled to oats, and pluck the heads very expertly,

* Schwenckfeld.

† The *Betula Nana*, or *Dwarf Birch*, a native of the arctic regions.

with green-peafe, hemp-feed, millet, the feeds of dodder*; but hemp-feed fattens them too faft, and occafions their death.

They return in fpring to their icy fummits. Though they do not always hold the fame route, they are commonly feen in Sweden, in Saxony, in Lower Silefia, in Poland, in Red Ruffia, in Podolia, and in Yorkfhire†. They are very rare in the fouth of Germany, and almoft entirely unknown in Switzerland and Italy‡.

In the time of their paffage they keep conftantly along the roads, picking up fmall feeds, and every fort of food. This is the proper feafon for catching them. They are prized on account of the fingularity of their plumage, and the delicacy of their flefh, and not for the fake of their fong, which confifts in an unmeaning cluck, or in a fhrill cry refembling that of the Jay, which they utter when handled: but to judge fairly of their warble, we ought to hear them in the feafon of love, when the warmth of paffion infpires, and foftens the native ftrains. We are ignorant alfo of the particulars of their breeding: it is undoubtedly in the countries where they fpend the fummer, but there are not many obfervers in the Lapland Alps.

* *Cufcuta Europæa.* LINN.

† Willughby killed one in Lincolnfhire. Ray fays that numbers of them are caught during winter in Yorkfhire. Johnfon informed Willughby that fome are feen on the chain of the Northumbrian mountains.

‡ Gefner and Aldrovandus.

These

These birds do not perch; they continue always on the ground, where they run and trip about like our Larks, to which they are similar also in their port, in their size, in their long spurs, &c. but differ in the shape of the bill and tongue, in their plumage, in their migrations, in their arctic abodes, &c.*

It is observed that they sleep little or none in the night, and begin to hop by the earliest dawn. Perhaps this is the reason why they prefer the lofty mountains of the north in summer, where the day lasts the whole season.

Total length six inches and a half; the bill five lines; on the palate is a small tubercle that distinguishes the genus; the hind-toe is equal to that of the middle, and the nail is much longer, and less hooked; the alar extent eleven inches and one fourth; the tail two inches and two-thirds, somewhat forked, consisting of twelve quills, and projecting ten lines beyond the wings. [A]

* Some able naturalists have ranged the Snow Ortolan with the Larks; but Linnæus has with great propriety referred it to the Buntings.

[A] Specific character of the Snow Bunting, *Emberiza Nivalis*:—"Its wing-quills are white, the primaries black without; the "tail-quills black, the three lateral white." These birds are of the size of the Chaffinch. They probably breed in Spitzbergen, and certainly in Greenland, where they make their nests in the fissures of the mountain rocks, employing grass for the outside, feathers for the inside, and the down of the arctic fox for lining. They lay five white eggs, spotted with brown. They sing sweetly, sitting on the ground.

In

In autumn the Snow Buntings iſſue from their frozen retreats, and viſit the northern parts of Europe. They appear in Sweden in ſevere weather, and are thence called *Ill-vars-vogel*, and *Hard-vars-fogel*; and many are caught near the ſea ſhores. They enter Britain by the north of Scotland; at firſt they are lean, but ſoon grow fat and delicious; the Highlands abound with them.

The Snow Buntings appear alſo at Hudſon's-bay in April, retire northward in May to breed, and pay another viſit in September. They live in vaſt flocks, feed on graſs-ſeeds, &c. and are eaſily caught.

Their colour being produced by the degree of cold to which they are expoſed, is neceſſarily variable.

VARIETIES of the SNOW BUNTING.

It will be readily perceived from what we have ſaid in regard to the winter and ſummer dreſs of theſe birds, that we are not here to conſider the Varieties which belong to the two principal epochs, or the intermediate terms; theſe being only the ſhades which mark the progreſſive influence of cold or heat.

I. The JACOBINE BUNTING*. It is a variety of climate. Its bill, breaſt, and belly, are white; the legs gray, and all the reſt black. It appears every winter in Carolina and Virginia, and retires in the ſummer to breed probably in the north.

* This is the *Black Bunting* of Pennant and Latham, and the *Snow Bird* of Cateſby and Kalm; it is the *Emberiza Hyemalis* in the Linnæan ſyſtem, and the *Hortulanus Nivalis Niger* of Briſſon. Specific character:—"It is black, its belly white." Theſe birds breed in the northern parts of America, winter in the ſtate of New York, and in very ſevere ſeaſons viſit Virginia and the Carolinas in immenſe bodies. They frequent the gardens and hillocks, to pick up the ſcattered graſs-ſeeds. The Americans call them *Chuck-birds*, and eſteem them delicious eating.

II. The

II. The COLLARED SNOW BUNTING *. The head, throat, and neck, are white. It has two ſorts of collars at the under-part of the neck; the upper lead-colour, the lower blue; both ſeparated by the ground colour, which forms a ſort of white intermediate collar; the quills of the wings are white, tinctured with a greeniſh yellow, and ſtrewed with ſome black feathers; the eight quills of the middle of the tail, and the two outer ones, white; the two others black; all the reſt of the plumage reddiſh brown, ſpotted with greeniſh yellow; the bill red, edged with cinereous; the iris white, and the legs fleſh-coloured.—This bird was caught in the county of Eſſex; it could not be enſnared till after many and tedious trials.

Kramer obſerves that in the Ortolans, as well as in the Yellow Buntings, the Chaffinches, and Bulfinches, the two mandibles are moveable; and this is the reaſon, he ſays, why they ſhell the ſeeds, and do not ſwallow them entire.

* This is the *Pied Chaffinch* of Albin and Latham, the fourth variety of Linnæus's *Snow Bunting*, the *Fringilla Capite Albo* of Klein, and the *Hortulanus Nivalis Torquatus* of Briſſon.

The RICE BUNTING.

L'Agripenne, ou *L'Ortolan de Riz,* Buff.
Emberiza Oryzivora, Linn. and Gmel.
Hortulanus Carolinenſis, Briſſ.
Emberiza Carolinenſis, Klein.
The Rice Bird, Cateſby.

THESE birds are migratory birds, and the motive of their paſſage is not known. Numerous flocks of them are ſeen or rather heard in the month of September, coming from the Iſland of Cuba, where the rice has already attained maturity, and directing their courſe to Carolina, where it is only coming into ear. Theſe remain in Carolina only three weeks, and then advance towards the north, always in ſearch of more tender grain: and, by ſucceſſive ſtations, they penetrate as far as Canada, and perhaps beyond. But what is the moſt ſingular, though there are other ſimilar inſtances, theſe flocks are compoſed entirely of females. It is aſcertained, we are told, from numerous diſſections, that only the females paſs in September; but in the beginning of the ſpring, the males and females are intermingled; and indeed, this is the ſeaſon of the union of the ſexes.

The plumage of the females is ruſty over almoſt its whole body; that of the males is more diverſified:

diversified: the fore-part of the head and neck, the throat, the breast, and all the under-part of the body, the upper part of the back and the thighs, black, with a mixture of rusty; the back of the head and neck rusty; the lower-part of the back and rump of an olive cinereous; the great superior coverts of the wings of the same colour, edged with whitish; the small superior coverts of the wings, and the superior coverts of the tail, dirty white; the quills of the wing black, tipt with brown, and edged, the great ones with sulphur-colour, and the small ones with gray; the quills of the tail are nearly like the great quills of the wings, only all terminate in points *; lastly, the bill is cinereous, and the legs brown. This Ortolan is remarked to be taller than the rest.

Total length six inches and three-fourths; the bill six lines and a half; the alar extent four inches; the tail two inches and a half, somewhat forked, and exceeding the wings by ten lines. [A]

* For this reason we have called the bird *Agripenne*.

[A] Specific character of the Rice Bunting, *Emberiza Oryzivora*:—"Black, the neck tawny, the belly black, the tail-quills "pointed." It is remarkable that the Rice-birds were not known in Carolina before the end of last century, when that nutritious grain was introduced from Madagascar. We may suppose that a few stray birds had been driven into that province by adverse winds, and had fared so well among the rice-crops, as to have returned with their brood the ensuing season: and thus, in the course of a few years, a direction would be given to their general migrations. They arrive in Carolina about September; at first they are

are very lean, but soon grow excessively fat, and fly with difficulty, so that they are easily shot. Their stay lasts three weeks; and both sexes make a transient visit in the spring. A few remain through the winter in Carolina, and even in Virginia, where they subsist on the scattered grains of Indian corn.

VARIETIES *of the* RICE BUNTING.

The LOUISIANA BUNTING.

L'Agrippenne, ou *Ortolan de la Louisiane*, Buff.
Emberiza Oryzivora, Var. Linn.

I consider this as a variety of the preceding, produced by the influence of climate: its size, its port, its shape are the same, and the quills of the tail are likewise pointed, the only difference, in short, consisting of the colours of the plumage. The Louisiana Bunting has the throat and all the under-part of the body of a light yellow, which is still more dilute on the lower-belly; the upper-part of the head and of the body, and the small superior coverts of the wings, are of an olive-brown; the rump and the superior coverts of the tail, yellow, finely striped with brown: the quills of the tail blackish, those of the middle edged with yellow, the lateral ones with white, the intermediate ones with the different shades that intervene between white and yellow; the great superior coverts of the wings are

are black, edged with white ; the quills are the ſame, except the middle ones, which have more white.

The meaſures are nearly the ſame as in the Rice Bunting *.

* Mr. Pennant thinks that this bird is the female of the common Rice Bunting.

The YELLOW BUNTING*.

Le Bruant de France, Buff.
Embèriza Citrinella, Linn. Gmel. Scop. Will. Kram. &c.
Emberiza Flava, Gesner.
Emberiza, Briss.
The Yellow Youlring, Sibbaldi Scotia Illustrata.
The Yellow Hammer, Ray.

THE osseous tubercle or barley-corn on the palate of this bird, proves indisputably its affinity to the Ortolans: but it resembles them also by other properties; by the shape of its bill and tail, by its proportions, and by the delicacy of its flesh †. Salerne remarks, that its cry is nearly the same.

The Yellow Bunting makes several hatches, the last in September: it places its nest on the

* In Germany, it is called *Emmerling, Geel-ammer, Gerst-ammer, Gruen-zling, Gaelgensicken, Gilbling, Gilberschen, Gilwertsch, Korn-vogel, Geel-gorst*; which names allude in general to its yellow plumage and its feeding upon corn, especially barley; in Switzerland, *Emmeritz, Embritz, Emmering, Hemmerling*; in Italy, *Zivolo, Zigolo, Cia Megliarina, Vetzero, Paietzero, Spaiarda*; in Brabant, *Jasine*; in Illyria, *Struad*; in Sweden, *Groening*; in Denmark, *Gulspury, Gulvesling*; in Smoland, *Golspinck*. In Latin, it was termed *Galgulus*, or *Galbula*, and also *Icterus*, from the Greek Ικτερος, signifying the jaundice, both on account of its yellow plumage, and a notion entertained by the people that the sight of it cured that disease. PLIN. xxx. 11.

† Its flesh is yellow, and has been said to be a remedy for the jaundice: nay, a person afflicted by that disease might transfer it by looking at the bird. SCHWENCKFELD.

ground,

N.º 102

THE YELLOW HAMMER.

ground, below a clod, in a buſh, or in a tuft of graſs, but always careleſsly. Sometimes it builds in the low branches of ſhrubs, and is then at more pains. The body of the neſt conſiſts of ſtraws, moſs, and dry leaves, and is lined with roots, the fineſt ſtraws, hair, and wool. The eggs are generally four or five, ſpotted with brown of different ſhade, on a white ground; but the ſpots are thicker at the large end. The female covers with ſuch ardent attachment, that often ſhe can be caught by the hand in broad day. The young are fed with ſmall ſeeds, inſects, and even May-flies; but of theſe laſt, the hard cruſt ſheathing the wings is previouſly ſeparated by the parents. They are however granivorous, and fondeſt of millet and hemp-ſeed. They can be caught by a nooſe baited with a head of oats; but cannot be decoyed, it is ſaid, by the call. In ſummer they haunt the trees, the ſides of the hedges and buſhes; ſometimes they viſit the vineyards, but ſcarcely ever penetrate into the heart of the foreſts. In winter, a part of them migrate into other climates, and thoſe which remain behind aſſemble and join the Chaffinches, Sparrows, &c. forming very numerous flocks, eſpecially in rainy weather. They reſort to the farms, and even to the villages and high roads, picking up their ſubſiſtence among the buſhes, and even in horſe-dung, &c. and in that ſeaſon they are almoſt as familiar

 as

as the Sparrows *. They fly rapidly and alight ſuddenly, and for the moſt part in the midſt of the thickeſt foliage, and never upon a ſeparate branch. Their ordinary cry conſiſts of ſeven notes, ſix of which are equal, and of the ſame tone, and the laſt ſharper and prolonged, *tĭ, tĭ, tĭ, tĭ, tĭ, tĭ, tī* †.

The Yellow Buntings are ſpread over the whole of Europe, from Sweden to Italy, and through all the interjacent countries; and they are conſequently expoſed to great difference of temperature, which happens to moſt birds in any degree domeſtic.

The male is diſtinguiſhed by the bright yellow feathers on the head and on the lower-part of the body; but on the head this colour is variegated with brown; it is pure yellow on the ſides of the head, under the throat, under the belly, and on the inferior coverts of the wing

* Friſch derives the German name *Ammer*, or *Hammer*, from *ham*, which ſignifies a houſe: *Ammer*, on this hypotheſis, would denote *domeſtic*. [The old Engliſh name *Yellow Hammer* is evidently borrowed from the German.]

† According to ſome, they have another cry, *vignerot, vignerot, vignerot, titchye*. Olina ſays, that they partly imitate the warble of the Chaffinches, with which they aſſociate. Friſch relates that they adopt ſomething of the ſong of the Canary when they hear it young: he adds, that the croſs breed of the cock Bunting and hen Canary chants better than the father. Laſtly, Guys ſays, that the ſong of the cock Bunting grows pleaſant on the approach of the month of Auguſt. Aldrovandus alſo ſpeaks of its fine warble.

wings, and it is mixed with light chesnut on all the rest of the lower part; the neck and the small superior coverts of the wings are olive; the middle and large coverts of the wings, the back, and even the four first quills of the wings are blackish; the rest are brown, and edged, the two outer with white, and the ten others with whitish gray; lastly, their superior coverts are light chesnut, terminated with whitish gray. The female has not so much yellow as the male, and is more spotted on the neck, the breast, and the belly: in both, the edges of the lower mandible are received into the upper, whose edges are scalloped near the point; the tongue is divided at the tip into slender threads; and lastly, the hind claw is the longest of all. The bird weighs five or six gros; the intestinal tube is seven inches and a half long; vestiges of a *cæcum;* the *æsophagus* is two inches and a half long, dilating near the gizzard, which is muscular; the gall bladder very small. I found in the *ovarium* of the females which I dissected, eggs of unequal bulk.

Total length, six inches and one third; the bill five lines; the legs eight or nine lines, the middle toe almost as long; the alar extent nine inches and one-fourth; the tail two inches and three-fourths, consisting of twelve quills, somewhat forked, not only because the intermediate quills are shorter than the lateral ones, but also because the six quills on each side turn naturally

 outwards;

outwards; they extend twenty-one lines beyond the wings. [A]

[A] Specific character of the Yellow Bunting, *Emberiza-Citrinella*:—" Its tail-quills are blackiſh, the two outermoſt marked with " a ſharp white ſpot on the inſide." Thus deſcribed by Briſſon: " Above variegated with tawny-blackiſh and white-gray; below " yellowiſh; the breaſt variegated with dilute cheſnut, yellowiſh, " and olive; the head yellowiſh, varied with duſky ſpots; a duſky " bar behind the eyes; the two outmoſt tail-quills on both ſides " marked within with a white ſpot." It is ſix inches and a half long; very frequent in England. It lays ſix eggs, which are whitiſh-purple, with blackiſh irregular ſpots and ſtreaks.

VARIETIES *of the* YELLOW BUNTING.

The colours vary, in different ſubjects and in different climates, both their ſhades and diſtribution: ſometimes the yellow extends over all the head, neck, &c. In ſome, the head is of a yellowiſh cinereous; in others the neck is cinereous, ſpotted with black; the belly, the thighs, and the legs are ſaffron-colour; the tail brown, edged with yellow, &c.

The CIRL BUNTING.

Le Zizi, ou *Bruant de Haie* *, Buff.
Emberiza Cirlus, Linn. and Gmel.
Emberiza Sepiaria, Briss.
Cirlus, Aldrov.
The Cirlus, or *Zivolo*, Will.

THIS is seen sometimes perched, sometimes running on the ground, and particularly in newly ploughed fields, where it finds seeds, small worms, and other insects; and accordingly it almost always has earth sticking to its bill. It is easily ensnared, and when caught with bird-lime, it ofteneſt remains attached, or if it entangles itself, it loses most of its feathers in the struggle, and is no longer able to fly. It soon becomes reconciled to captivity, but is not absolutely insensible of its situation; for, during the first two or three months, it has only its usual chirp, which it repeats with frequency and trepidation when a person goes near its cage: however, by gentle treatment, it at length resumes its warble. Its size and its habits are nearly the same with those of the Yellow Bunting, and probably, if we were better acquainted with these birds, we should perceive that they belonged to the same species.

* i. e. The Hedge-Bunting.

The Cirl Buntings are not found in the northern countries, and ſeem to be moſt frequent in thoſe of the ſouth; however, they are rare in ſeveral provinces of France. They are often ſeen with the Chaffinches, whoſe ſong they imitate, and with whom they form numerous flocks, eſpecially in rainy days. They feed on the ſame ſubſtances as the other granivorous tribe, and live about ſix years according to Olina: but this muſt be underſtood of them in the domeſtic ſtate; for it is uncertain what effects freſh air and freedom of motion may have upon longevity.

In the male, the upper-part of the head is ſpotted with blackiſh upon an olive-green ground; there is a yellow ſpot on the ſides, divided into two unequal parts, by a black ſtreak which paſſes over the eyes; the throat is brown and alſo the top of the breaſt, and a yellow collar lies between them; the reſt of the under-part of the body is yellow, which grows more dilute as it ſpreads to the tail, and is ſpotted with brown on the flanks; the upper-part of the neck and back is variegated with rufous and blackiſh; the rump olive-rufous, and the ſuperior coverts are of a purer rufous; the quills of the wings brown, edged with olive, except the neareſt to the back, which are rufous, and the two middle ones, which are ruſty-gray; laſtly, the bill is cinereous, and the legs brown.

In

In the female, there is leſs of the yellow, the throat is not brown, nor does any brown ſpot appear on the breaſt.

Aldrovandus tells us, that the plumage is ſubject to much diverſity in this ſpecies: the one which he figured had a dull green tinge on its breaſt; and of thoſe which I have obſerved, I found one of which the upper-part of the neck was olive, with ſcarcely any admixture.

Total length ſix inches and one-fourth; the bill about ſix lines; the alar extent nine inches and three-fourths; the tail near three inches, compoſed of twelve quills, and projecting about ten lines beyond the wings, and forked as in the Yellow Bunting. [A]

[A] Specific character of the Cirl-Bunting, *Emberiza-Cirlus*:—"It is brown, its breaſt ſpotted, its eye-brow yellow, the two outermoſt quills of its tail marked with a white wedge-ſhaped ſpot."

M

The FOOLISH BUNTING.

Le Bruant Fou, Buff.
Emberiza-Cia, Linn. Gmel. Kram.
Emberiza Pratensis, Briss. and Gesner
Emberiza Barbata, Scop.
Cirlus Stultus, Ray, and Will.

THE Italians have applied the epithet of *Foolish* to this bird, on account of its incautious disposition, being readily caught in every sort of snare: but the want of circumspection is characteristic of the genus, and the Foolish Bunting is inconsiderate only in a higher degree. The name of *Meadow Bunting* is improper; for the most observant bird-catchers and fowlers have unanimously assured me, that they never saw it in the meadows.

Like the Cirl Bunting, the Foolish Bunting is not found in the northern countries, nor does its name occur in the catalogues of the Swedish and Danish birds. It prefers solitude, and delights in mountainous abodes. It is very common and well known in the hills round Nantua. Hebert * often saw it on the ground, and upon the chesnut-trees; and the country people told him, that its flesh was excellent meat. Its song is very ordinary, and resembles that of the Yellow Bunting; and the Prussian bird-catchers

* This excellent observer has communicated or confirmed the principal facts of the history of the Buntings.

have

have remarked, that when it is put into a volery among others of a different ſpecies, it diſcovers a ſtrong predilection for the Yellow Bunting. Indeed its cry*, its ſize, its figure, are the ſame, and it differs only by ſome of its habits.

In the male, all the upper-part is variegated with blackiſh, and gray; but this gray is purer on the head, and ruſty every-where elſe, except on ſome of the middle coverts of the wings, where it becomes almoſt white. The ſame ruſty gray edges almoſt all the quills of the wings and of the tail, whoſe ground colour is brown, only the two exterior quills of the tail are edged and tipt with white; the orbits are ruſty white; the ſides of the head and of the neck are gray; the throat is gray, dotted with blackiſh, and edged on each ſide and below by a line almoſt black, which forms an irregular ſort of ſquare with the gray plate on the ſides of the head; all the under-part of the body is fulvous, more or leſs dilute, but dotted or variegated with blackiſh on the throat, the breaſt, and the flanks; the bill and legs are gray.

Total length ſix inches and one fourth; the bill five or ſix lines; the alar extent nine or ten inches; the tail two inches and one third, a little forked, conſiſting of twelve quills, and exceeding the wings by ſixteen lines. [A]

* Linnæus ſays that in flying it chirps *zip, zip*.

[A] Specific character of the Fooliſh Bunting, *Emberiza-Cia*:—
"It is ruſty, its head marked with ſcattered blackiſh lines, its eye-
"brows white."

The COMMON BUNTING*.

Le Proyer, Buff.
Emberiza Miliaria, Linn. Gmel. Kram. and Frifch.
Emberiza Alba, Will. and Klein.
Cynchramus, Briff.

THIS is a bird of paffage, and arrives early in the fpring. It deferves to be called *Meadow Bunting*, fince in the fummer feafon it never ftrays far from the low grounds †. It makes its neft among the fields of barley, of oats, of millet, &c. feldom on the furface of the ground, but three or four inches above, among the thick ftrong herbage ‡. The female lays four, five, and fometimes fix eggs; and while fhe is engaged in hatching, the male brings her food, and fitting on the fummit of a tree, he repeats inceffantly the difagreeable cry, *tri*, *tri*, *tri*, *tiritz*, which he retains only till the month of Auguft: the notes are fharper and fhorter than thofe of the Yellow Bunting.

It has been obferved, that when the Bunting rofe from the ground towards a branch, its legs dropped, and its wings quivered with an irre-

* In German, *Knuft*, *Knipper*, *Gerft-Ammer* (Barley-Bunting), *Graue-Ammer* (Gray Bunting); in Swedifh, *Korn-laerka* (Corn Lark); in Norwegian, *Knotter*. The Italian name *Strilozzo* comes from *Strillare*, to *creck*, on account of its cry. In Greek it was called Κυγχραμος, or Κυχςανος, according to Belon.

† Belon fays that it follows the water like the Woodcock.

‡ Belon.

gular

Nº 103

THE BUNTING.

gular motion peculiar to the ſeaſon of love. At other times, in autumn for inſtance, it flies equably and ſwiftly, and mounts to a conſiderable height.

The young ones leave the neſt long before they are able to fly, and take delight to run among the graſs; and this would ſeem to be the reaſon why the parents build ſo cloſe to the ground. The pointer-dogs often ſurpriſe them in the chace of quails. The parents ſtill continue to feed and guard them till they are fledged; but their anxiety for the ſafety of their brood often betrays them, and if a perſon chances to go near the ſpot, they circle his head with a doleful air.

After the family is raiſed, they pour their numerous flocks into the fields, eſpecially among the crops of oats, beans, and the late ſorts of grain. They migrate ſoon after the Swallows, and it very ſeldom happens that any of them remain during the winter *.

It is obſerved that the Bunting does not flutter from branch to branch, but alights on the extremity of the higheſt and moſt detached bough either of a tree or ſhrub, and in a moment begins its ſong, which it prolongs for whole hours in the ſame place, repeating its tireſome note, *tri*, *tri*; and laſtly, that in taking flight, it chatters with its bill †.

* Geſner.

† Moſt of theſe facts were communicated by M. Hebert.

The female ſings alſo, after the young no longer occupy her attention; but this is only when perched on a branch, and about mid-day. Her ſong is as bad as that of the male. She is rather ſmaller, but her plumage is nearly the ſame. Both feed upon grain and ſmall inſects, which they find in the fields and meadows.

Theſe birds are ſpread over all Europe, or rather they viſit the whole extent of it in their migrations. Olina affirms that they are more numerous at Rome and in its vicinity than in other countries. Bird-catchers keep them in a cage to uſe as calls in autumn; and they not only entice the Fooliſh Buntings into the ſnare, but many other ſmall birds of different kinds. They are for this purpoſe put in low cages without any bars or rooſts.

In the male, the upper-part of the head and body is variegated with brown and rufous; the throat, and the orbits, light rufous; the breaſt, and all the reſt of the under-part of the body, yellowiſh-white, ſpotted with brown on the breaſt and ſides; the ſuperior coverts of the wings, their quills, and thoſe of the tail, are brown, edged with rufous, more or leſs dilute; the bill and legs brown gray.

In the female, the rump is gray, verging upon rufous, without any ſpots; the ſuperior coverts of the tail the ſame colour, edged with whitiſh; and in general the quills of the tail, and of the wings, are bordered with lighter colours.

The bill of these birds is of a remarkable shape; the two mandibles are moveable, as in the Ortolans; the edges are also re-entering, as in the Common Bunting, and the junction is made in a crooked line; the edge of the lower mandible on each side, near the third of its length, makes an obtuse salient angle, and is received by the corresponding *re-entrant* angle in the upper-mandible, which is more solid and bulky than in most other birds; the tongue is narrow, thick, and tapered to a point, like a tooth-pick; the nostrils are covered above by a membrane of a crescent shape, and below by small feathers; the first *phalanx* of the outer toe is joined to that of the middle toe.

Intestinal tube thirteen inches and a half; the gizzard muscular, preceded by a moderate dilatation of the *æsophagus*, containing vegetable matter, and nuts with small pebbles; slight vestiges of a *cæcum;* no gall-bladder; the great axis of the testicles four lines, the smaller one three lines. Total length of the bird seven inches and a half; the bill seven lines; the alar extent eleven inches and one third; the tail nearly three inches, somewhat forked, consisting of twelve quills, and stretching eighteen lines beyond the wings. [A]

[A] Specific character of the Common Bunting, *Emberiza Miliaria*;—" It is brown, below spotted with black its orbits rufous.

FOREIGN BIRDS,

WHICH ARE RELATED TO THE BUNTINGS.

I.

The BRAZILIAN BUNTING.

Le Guirnegat, Buff.
Emberiza Brasiliensis, Gmel. and Briss.

HAD not this bird been a native of South America, and its cry been different from that of the Yellow Bunting *, I should have considered it as a mere variety. Indeed its plumage has even more of the Yellow than common in ours †, and I have no doubt but they would intermix, and beget prolific offspring.

The yellow is spread unmixed on the head, the neck, and all the under-part of the body, and also borders almost all the superior coverts, and the quills of the tail and of the wings, which are brown; on the back it is intermingled with brown and green; the bill and the eyes are black, and the legs brown.

* Our Bunting is called *Luteola, Aureola, Gold-hammer, Bruant Jaune, Bruant Doré, Cia Pagliarina:* so that yellow would seem to form part of its essence.

† Some individuals of our Bunting have the head, the neck, and the upper-side of the body almost entirely yellow; but this is rare.

This

This bird is found in Brazil, and probably is indigenous, for the natives have given it a name, *Guiranbeemgata*. Marcgrave praiſes its ſong, and compares it to that of the Chaffinch.

The female is very different from the male; for the ſame author tells us, that the plumage and cry reſemble thoſe of the Sparrow. [A]

[A] Specific character of the *Emberiza Braſilienſis*: "Its top, its "neck, and the under-ſide of its body, yellow; its back, its wings, "and its tail, variegated with yellow and brown."

M

II.

The MEXICAN BUNTING.

La Thereſe Jaune, Buff.
Emberiza Mexicana, Gmel.

As I have ſeen only the figure, and a dead ſpecimen of this bird, I can give but an imperfect deſcription. Its plumage is much like that of the Common Yellow Bunting; almoſt all the head, the throat, and the ſides of the neck, are orange-yellow; the breaſt, and the under-part of the body, ſpeckled with brown on a dirty white ground; the back of the head, and neck, and all the upper-part of the body, brown: this laſt colour tapers to a point on each ſide of the neck, and extends almoſt to the eye. The quills of the wings and tail, and their coverts, are brown, edged with a lighter brown. [A]

[A] Specific character of the *Emberiza Mexicana*:—"Above it "is ruſty; below partly white, ſpotted with brown; its head, and "throat, yellow."

 III. The

III.
The YELLOW-FACED BUNTING.

La Flaveole, Buff.
Emberiza Flaveola, Linn. and Gmel.

The forehead and throat are yellow, and all the reſt of the plumage gray. It is nearly of the ſize of the Siſkin. Linnæus, who has made us acquainted with this ſpecies, informs us, that it is a native of the warm countries, but does not mention to what continent it belongs. [A]

[A] Specific character of the *Emberiza Flaveola*:—" It is gray; " its face yellow."

IV.
The OLIVE BUNTING.

L'Olive, Buff.
Emberiza Olivacea, Linn. and Gmel.
Emberiza Dominicenſis, Briſſ.

This little Bunting, which is found in Dominica, exceeds not the ſize of a Wren. All the upper-part, and even the tail, and the quills of the wings, are of an olive-green; the throat orange-yellow; there is a ſpot of the ſame colour between the bill and the eye; the fore-part of the neck is blackiſh; all the under-part of the body a very light gray, tinged with olive; the anterior part of the wings edged with light yellow; the bill and legs brown.

The female has not the black neck-piece, nor the orange yellow ſpot between the bill and the eye; nor is the throat orange-yellow, as in the male.

Total length three inches and three-fourths; the bill four lines and a half; the alar extent ſix inches; the tail eighteen lines, conſiſting of twelve quills, and projecting ſeven or eight lines beyond the wings. [A]

[A] Specific character of the *Emberiza Olivacea*: — "It is "olive; waiter below; its throat orange; a ſtripe on its breaſt "blackiſh."

V.

The AMAZON BUNTING.

L'Amazone, Buff.
Emberiza Amazona, Gmel.

This bird was found at Surinam. It is of the bulk of our Titmouſe; the upper-part of the head is fulvous; the inferior coverts of the wings whitiſh; the reſt of the plumage brown.

VI.

The PLATA BUNTING.

L'Emberiſe à Cinq Couleurs, Buff.
Emberiza Platenſis, Gmel.

This bird was brought from Buenos-Ayres. We deſcribe it on the authority of Commerſon, who ſpeaks only of its plumage and external

 cha-

charaсters, and takes no notice of its manner of living, nor informs us whether it has the discriminating properties of the ſpecies.

All the upper-part of the body is of a brown green, verging to yellow ; the head, and the upper-part of the tail, of a darker tinge ; the under-part of the tail has more of a yellow caſt the back marked with ſome black ſtreaks ; the anterior edge of the wings bright yellow ; the quills of the wings, and the outermoſt of thoſe of the tail, edged with yellowiſh ; the under-part of the body cinereous white ; the pupil blackiſh blue ; the iris cheſnut ; the bill cinereous, convex, and pointed ; the edges of the lower mandible *re-entrant ;* the noſtrils covered with a membrane very near the baſe of the bill ; the tongue terminating in ſmall filaments ; the legs lead-coloured.

Total length eight inches ; the bill eight lines ; the alar extent ten inches ; the tail four inches ; the hind nail largeſt of all.

M

VII.

The BOURBON BUNTING.

Le Mordoré, Buff.
Emberiza Borbonica, Gmel.

The whole of the body is reddiſh gray, both above and below, and almoſt of the ſame ſhade ; the coverts of the wings, their quills, and thoſe

 of

of the tail, are brown, edged with reddiſh gray, more or leſs dilute; the bill brown, and the legs yellowiſh, tinged ſlightly with reddiſh gray. It is found in the Iſle of Bourbon, is nearly of the bulk of the Yellow Hammer, but its tail is ſhorter, and its wings longer; the former projecting about ten lines beyond the latter.

VIII.

The GRAY BUNTING.

Le Gonambouch, Buff. and Seba.
Emberiza Griſea, Gmel.
Emberiza Surinamenſis, Briſſ.

Seba tells us that this bird is very common at Surinam, that it is of the ſize of the Lark, and that it ſings like the Nightingale, and conſequently much better than any of the Buntings; which is extraordinary in an American bird. The people of the country ſay, that it is extremely fond of maize, and often perches upon the top of the ſtalk.

Its principal colour is light gray, but there is a tinge of red on the breaſt, the tail, the coverts, and the quills of the wings; the quills of the wings are white below.

Total length five inches; the bill five lines; the tail eighteen lines, and exceeding the wings by ten lines.

 IX. The

IX.

The FAMILIAR BUNTING.

Le Bruant Familier, Buff.
Emberiza Familiaris, Linn. and Gmel.
Motacilla Familiaris, Osb. It.

The head and bill are black; the upper-part of the body cinereous and spotted with white; the under-part cinereous, but without spots; the rump and part of the back that is covered by the wings, yellow; the coverts and the end of the tail, white.—This bird is found in Asia, and is nearly the size of the Siskin.

X.

The CINEREOUS BUNTING.

Le Cul-Rousset, Buff.
Emberiza Cinerea, Gmel.
Emberiza Canadensis, Briss.

We are indebted to Brisson for this species:—the upper-part of the head variegated with brown and chesnut; the under-part of the neck, the back, and the coverts of the wings, variegated in the same manner with a mixture of gray; the rump gray without spots; the superior and inferior coverts of the tail, dirty white and rusty; the

the throat and all the under-part of the body, dirty-white, variegated with chefnut fpots, lefs frequent however below the belly; the quills of the tail and of the wings brown, edged with gray, verging upon chefnut; the bill and legs brown-gray.—It was brought from Canada.

Total length five inches and a half; the bill five lines and a half; the alar extent eight inches and one-fourth; the tail two inches and a half, confifting of twelve quills, and projecting about twenty-one lines beyond the wings.

XI.

The BLUE BUNTING.

L'Azuroux, Buff.
Emberiza Cærulea, Gmel.
Emberiza Canadenfis Cærulea, Briff.

We are alfo indebted to Briffon for this Canadian bird: the upper-part of the head is dull rufous; the upper-part of the neck and of the body variegated with the fame and with blue; the rufous is not fo deep on the fmall coverts of the wings nor on the large ones, which are edged and tipt with that colour; the quills of the wings and of the tail are brown, edged with blue-gray; the bill and legs are brown-gray.

Total length four inches and one-fourth; the bill five lines; the alar extent feven inches and one-third; the tail an inch, consisting of twelve quills, and not exceeding the wings by more than four lines.

XII.

The BONJOUR COMMANDER.

This is the name which the settlers in Cayenne give to a kind of Bunting, which frequents the dwellings and sings at day-break. Some call it the Cayenne Bunting. It resembles the one from the Cape of Good Hope so exactly, that Sonini thinks it is the same. One of the appellations ought therefore to be rejected; and this shews that all epithets of birds that are geographical are insufficient to discriminate them.

The cry is shriller than that of our Sparrows; they are generally on the ground, and like the Buntings, they are for the most part in pairs.

The male has a black hood crossed by a gray bar; the cheeks are cinereous; there is a black ray extending from the base of the bill to the hood, and below and behind it there is a rufous half-collar; the upper-part of the body is greenish-brown, variegated on the back with oblong black spots; the coverts of the wings are edged with rusty; all the under-part of the body is cinereous.

It

It is a little ſmaller than the Cirl Bunting, its total length being only five inches; its wings are ſhort, and ſcarcely reach to the middle of the tail.

XIII.

The RED-EYED BUNTING.

Commerſon deſcribed this bird on the Iſle of France, which it inhabits, and where it is called *Calſat*. The upper part of the head is black, and all the upper-part of the body, including the wings and the tail, are bluiſh cinereous; the tail edged with black; the throat black; the breaſt and belly wine-coloured; a white bar ſtretches from the corner of the opening of the bill to the back of the head; the orbit of the eyes is naked, and roſe-coloured; the iris, the bill, and the legs alſo roſe-coloured; the inferior coverts of the tail white.

It is of a middle ſize, between the Sparrow and the Linnet.

The BULFINCH*.

Le Bouvreuil, Buff.
Loxia-Pyrrhula, Linn. and Gmel.
Coccothraustes Sanguinea, Klein.
Rubicilla, seu *Pyrrhula*, Aldrov. Johnst. &c.
Pyrrhula, Briss.
Rubrica, Gesner.
The Bulfinch, *Alp*, or *Nope*, Will.

NATURE has been liberal to this bird, for she has bestowed upon it a beautiful plumage and a fine voice. The colours are perfect after the first moulting, but the song needs to be assisted and formed by art. In the state of freedom, the Bulfinch has three cries, which are all unpleasant: the first, which is the most common, is a sort of whoop; it begins with one, then two in succession, and afterwards three and four, &c. and, when animated, it seems to articulate with force the repeated syllable *tui*, *tui*, *tui*; the second is an air of greater extent, but lower, almost hoarse,

* In German, *Blut-finch*, *Guegger*, *Brom-meiss*, *Bollen-beisser*, *Rot-vogel*, *Thumbherz*, *Gumpel*; in Swedish, *Dom-herre*; in Danish and Norwegian, *Dom-papp*, *Blod-finke*; in Polish, *Popek*; in Prussian, *Daun pfaffe*; in Italian, *Cifolotto*, *Suffuleno*, *Fringuello Montano*, *Fringuello Vernino*, *Monachino*. In Greek, it was called Συκαλις, from Συκον, a fig, on which it was supposed to feed; and also Πυρρυλας, from Πυρ, fire, on account of its red plumage.

and

N.º 104

THE BULFINCH.

and running into a diſcord *; and the third is a feeble ſtifled cry, which it vents at intervals, exceedingly ſhrill and broken, but at the ſame time ſo ſoft and delicate, that it ſcarcely can be heard; it emits this ſound much in the ſame way as a ventriloquiſt, without any apparent motion of the bill or throat, only with a ſenſible action of the abdominal muſcles.—Such is the ſong of the Bulfinch when left to the education of its parents; but if man deigns to inſtruct it methodically †, and accuſtom it to finer, mellower, and more lengthened ſtrains, it will liſten with attention, and the docile bird, whether male or female ‡, without relinquiſhing its nativc airs, will imitate exactly, and ſometimes ſurpaſs its maſter §. It alſo learns eaſily to articulate words and

* This is its warble, *sĭ, ŭt, ŭt, ŭt, ŭt, sĭ, rĕ, ŭt, ŭt, ŭt, ūt, ŭt, ŭt, sĭ, rĕ, ŭt.* With the ſame voice it alſo pronounced *ut, la, ut, mi, ut, la.* Sometimes theſe paſſages were preceded by a drawling tone, in the ſame ſtyle, but without any inflexion, and which reſembled a ſort of mewing. [The notes of the French gamut are Sol, La, Si, Ut, Re, Mi, Fa, correſponding to the Engliſh C, D, E, F, G, A, B.]

† It is ſaid, that to ſucceed in teaching the Bulfinches one ſhould whiſtle to them, not with the Canary-flageolet, but with the lipped or German-flute, whoſe tone is deeper and fuller. The Bulfinch can alſo mimic the warble of other birds.

‡ The hen Bulfinch is the only female, it is ſaid, of the ſinging-birds that learns to whiſtle as well as the male. *Ædonologie*, p. 18.—Olina.—Aldrovandus, &c. Some pretend that her voice is weaker and ſweeter than that of the cock Bulfinch.

§ " I know a curious perſon (ſays the author of the *Ædonologie*, p. 89.) who having whiſtled ſome airs quite plain to a Bulfinch, was agreeably ſurprized to ſee the bird add ſuch graceful turns, that the maſter could hardly recognize his own muſic, and acknowledged that the ſcholar excelled him." However, it muſt be confeſſed,

and phrafes, and utters them with fo tender an accent, that we might almoft fuppofe it felt their force.—The Bulfinch is befides fufceptible of perfonal attachment, which is often ftrong and durable. Some have been known, after efcaping from the volery and living a whole year in the woods, to recognize the voice of the miftrefs, and return, to forfake her no more *. Others have died of melancholy, on being removed from the firft object of their attachment †. Thefe birds well remember injuries received: a Bulfinch, which had been thrown to the ground in its cage by fome of the rabble, though it did not appear much affected at the time, fell into convulfions ever after at the fight of any mean looking fellow, and expired in one of thefe fits eight months from the date of its firft accident.

The Bulfinches fpend the fummer in the woods or on the mountains: they make their neft in the bufhes, five or fix feet from the ground, and fometimes lower: this confifts of mofs, lined with foft materials; and its opening is faid to be the leaft expofed to the prevailing wind. The female lays from four to fix eggs of

fefled, that if the Bulfinch be ill-directed, it acquires harfh ftrains. Hebert faw one which never had heard any perfon whiftle but carters, and which whiftled like them, with the fame ftrength and coarfenefs.

* One of thefe birds which returned to its miftrefs, after living a year in the woods, had all its feathers ruffled and tangled. Liberty has its inconveniences, efpecially for an animal depraved by domeftication.

† Ædonologie, p. 128.

of a dirty white and a little bluiſh, encircled near the large end with a zone, formed by ſpots of two colours, ſome of an ill-defined violet, others of a diſtinct black. She diſgorges the food for the young like the Goldfinches, the Linnets, &c. The male is attentive to his mate, and Linnæus relates that he ſometimes holds out to her a ſpider in his bill a very long time. The young ones begin not to whiſtle till they are able to eat without aſſiſtance; and then they ſeem inſtinctively benevolent, if what is related be true, that in a hatch of four, the three elder will feed their puny brother. After the breeding is over, the parents ſtill continue aſſociated through the winter, for they are always ſeen in pairs: thoſe which remain in the country, leave the foreſts, and deſcend from the mountains * when the ſnow falls, and forſake the vineyard which they haunt in the autumn, and approach our dwellings, or lodge among the hedges by the road-ſides: thoſe which migrate, depart with the Woodcocks, about All Saints day, and return in the month of April †. They feed in ſummer upon all ſorts of ſeeds, inſects, and ſorbs ‡; and in the winter, upon

* There are many Bulfinches in the mountains of Bologna, of Modena, of Savoy, of Dauphiné, of Provence, &c. OLINA.

† Many are ſeen about the end of autumn and the beginning of winter in the mountainous parts of Sueſia. but not every year, according to Schwenckfeld.

‡ Linnæus.

juniper-

juniper-berries, upon the buds of aſpen, of alder, of oak, of fruit-trees, of the marſh willow, &c. whence the name *Ebourgeonneux* (from *Bourgeon*), which they ſometimes have in France: in that forbidding ſeaſon, they are heard to whiſtle; and their ſong, though ſomewhat ſad, cheers the torpid gloom of nature *.

Some reckon theſe birds attentive and thoughtful; and their heavy air and the facility with which they are inſtructed ſeem to favour that idea; but, on the other hand, their allowing one to get near them, and their being decoyed into the different ſnares †, indicate want of circumſpection. As their ſkin is very tender, thoſe which are caught with bird lime loſe, in ſtruggling to eſcape, part of their down, and even of their quills, unleſs a perſon ſpeedily diſentangles them. It deſerves to be remarked, that thoſe which have the fineſt plumage are the leaſt capable of inſtruction, being older and not ſo tractable: but even the old ones are ſoon reconciled to the cage, provided that at firſt they have plenty of food; they can alſo be properly tamed, as I have already noticed, though to ſucceed requires time and patient attention, which is the reaſon that perſons ſometimes fail in the attempt. It ſeldom happens that one is caught alone; the

* In the cage they eat hemp-ſeed, biſcuit, prunes, ſallad, &c. Olina recommends for the young ones the Nightingale's paſte made with walnuts, &c.

† Geſner caught many of them during the winter, by a bait of night-ſhade berries.

other

other is ſoon enticed to follow its companion, and ſacrifices its freedom to the calls of friendſhip.

It has been aſſerted that the Canary, which breeds with ſo many other ſpecies, will never ſubmit to the embrace of the Bulfinch, and it is alleged as the reaſon, that the cock Bulfinch, when in heat, holds his bill open, which frightens the Canary. But the Marquis de Piolenc aſſures me, that he ſaw a Bulfinch pair with a hen Canary, which had five young ones about the beginning of April: their bill was larger than that of Canaries of the ſame age, and they began to be covered with a blackiſh down, which ſeemed to ſhew that they had more of the father than the mother: unfortunately they all died in performing a ſhort journey. What adds more weight to this obſervation, Friſch gives directions for the experiment: he adviſes that the cock Bulfinch be the ſmalleſt of its kind, and be kept long in the ſame volery with the hen Canary: he ſubjoins, that ſometimes a whole year elapſes before the female will allow the Bulfinch to come near, or to eat out of the ſame tray; which ſhews that the union is difficult but not impoſſible.

It has been obſerved that the Bulfinches jerk their tail briſkly upwards and downwards, though not in ſo remarkable a degree as the Wagtails. They live five or ſix years; their fleſh is palatable according to ſome, and not fit to be eaten according

according to others, by reaſon of its bitterneſs; this muſt depend upon the age, ſeaſon, and food. They are of the ſize of the Houſe-Sparrow, and weigh about one ounce. The upper-part of the head, the ring round the bill, and the origin of the neck, are fine gloſſy black, which extends more or leſs forwards and backwards *; the fore-part of the neck, the breaſt, and the top of the belly, beautiful red; the abdomen, and the inferior coverts of the tail and wings, white; the upper-part of the neck, the back, and the ſhoulders, cinereous; the rump white, the ſuperior coverts and the quills of the tail, fine black, verging to violet, a whitiſh ſpot on the outermoſt quills; the quills of the wings blackiſh cinereous, and deeper the nearer to the body: the laſt of all red on the outſide; the great coverts of the wings of a fine changing black, terminated with reddiſh light-gray; the middle ones cinereous; the ſmall ones, blackiſh aſh-colour, edged with reddiſh; the iris hazel; the bill blackiſh, and the legs brown.

The ſides of the head and the fore-part of the neck, the breaſt, the top of the belly, and in a word, almoſt all that was red in the male, is vinous aſh-colour in the female, and ſometimes even the abdomen: nor has it the fine gloſſy changing black that occurs on the head

* Hence the name of Monk, or Pope, which this bird has in many languages, and that of *Co'ally-hood,* given to it by the people in Scotland. T.

and

and other parts of the male. I have ſeen a female however, which had the laſt of the wing-quills edged with red, and which had no white on the outermoſt of thoſe of the tail. Linnæus adds, that the tip of the tongue is divided into ſmall filaments ; but I have always found it quite entire in the male, and ſhaped like a very ſhort tooth-pick.

In many young Bulfinches which I have obſerved about the end of June, the fore-head was light rufous, the fore-part of the neck and breaſt ruſty-brown ; the belly and the inferior coverts of the tail fulvous, which extends and grows more dilute on the ſide; the under-part of the body, more or leſs duſky ; the white ray over the wing deeply tinged with ruſty ; the rump white of different ſhades.—But it is obvious that conſiderable diverſities will occur.

Total length ſix inches; the bill five lines, thick and forked; Kramer remarks, that the two mandibles are moveable, as in the Finches and Buntings; the alar extent nine inches and one-fourth; the tail two inches and one-third, ſomewhat forked, (but not always in the females,) conſiſting of twelve quills; the outer-toe joined by its firſt *phalanx* to the mid-toe; the hind-nail ſtronger and more hooked than the reſt.

The dimenſions of the female when diſſected were as follows:—inteſtinal tube eighteen inches; veſtiges of a *cæcum* ; the *æſophagus* two inches

 and

and a half, dilated like a bag, with a projecting edge next the gizzard, which is muſcular, containing many ſmall pebbles, and even two or three ſmall yellow ſeeds quite entire, though the birds had remained two days and a half in the cage without eating; the cluſter of the *ovarium* of an ordinary ſize, and the ſmall eggs nearly equal; the ovi-duct ſpread, and above three inches in length; the *trachea* formed a ſort of knot of a conſiderable thickneſs, where it forked.

VARIETIES *of the* BULFINCH.

Sir Robert Sibbald has only a ſingle line on the Bulfinch, and ſays, that there are ſeveral kinds of it in Scotland: theſe are probably only the varieties which we ſhall now deſcribe.

Friſch tells that the Bulfinches may be divided into three different ſizes: the Marquis de Piolenc was acquainted with two diſtinguiſhed by their bulk*: and others pretend that thoſe of Nivernois are ſmaller than thoſe of Picardy. Lottinger aſſures us, that the Bulfinch of the mountains exceeds that of the plain; and this

* The ſmalleſt, adds M. de Piolenc, is of the ſize of the Chaffinch, its body is longer, its breaſt of a brighter red; and it ſeems wilder than the ordinary Bulfinch.

accounts

accounts for the diverſity of bulk, being variouſly modified by local ſituation. But theſe are too numerous to be ſeparately treated: and I ſhall here take notice only of the varieties of plumage.

I. The WHITE BULFINCH *. Schwenckfeld ſpeaks of a White Bulfinch that was caught near the village of Friſchbach in Sileſia, and which had only ſome black feathers on the back. This fact is confirmed by Deliſle: "There are "in this canton, (of Bereſow in Siberia,) ſays "that excellent aſtronomer, White Bulfinches, "whoſe back is ſomewhat blackiſh, and gray in "ſummer: theſe birds have a delicate pleaſant "ſong, much ſuperior to that of European "Bulfinches." It is probable that the northern climate has much contributed to this change of plumage.

II. The BLACK BULFINCH †. Under this denomination, I include not only thoſe which are entirely or almoſt black, but alſo thoſe which have perceptibly begun to aſſume that complexion:—ſuch was what I ſaw at Baron Goula's; its throat and rump were black; the inferior coverts of its tail, its lower belly, and the top of its breaſt, variegated with rufous wine-colour and black, and no white ſpots ap-

* *Loxia-Pyrrhula*, var. 2. Linn.
† *Loxia-Pyrrhula*, var. 1. Linn.

 peared

peared upon the laſt quill of the tail. Thoſe mentioned by Anderſon and Salerne were entirely of a jet black; that of Reaumur noticed by Briſſon, was black over the whole body. I have obſerved one which aſſumed a fine gloſſy black after the firſt moulting, but which ſtill retained a little red on each ſide of the neck, and a little gray behind the neck, and on the ſmall ſuperior coverts of the wings; its legs were fleſh-coloured, and the inſide of its bill red: that of Albin had ſome red feathers under the belly; the five firſt quills of the wing edged with white; the iris white, and the legs fleſh-coloured*. Albin remarks that this bird was exceedingly gentle, like all the Bulfinches. It often happens that this robe of black diſappears in moulting, and gives place to the natural colours: but often it renews each time, and remains for ſeveral years:—ſuch was the caſe with Reaumur's. This would imply that the change of colour is not the effect of diſeaſe.

III. The Great Black African Bulfinch†. Though this bird is a native of a diſtant country, and exceeds the ſize of the European Bulfinch, I am ſtill inclined to regard it

* Mr. White, in his Natural Hiſtory of Selburne, relates alſo an inſtance of a cock Bulfinch turning dingy, and afterwards black.

† *Loxia Panicivora*, Linn. and Gmel.
Pyrrhula Africana Nigra, Briſſ.
The White-winged Groſbeak, Lath.

it as analogous to the variety which I have defcribed by the name of the Black Bulfinch, and to fufpect that the burning climate of Africa communicates a black hue to the plumage, as the cold of Siberia introduces a fnowy white. It is entirely black except a very fmall white fpot on the great coverts of the wing; and alfo the bill is gray, and the legs afh-coloured. It was brought alive to Paris from the coafts of Africa.

Total length, feven inches and one-fourth; the bill fix lines; the alar extent four inches and one-fourth; the tail two inches and a half, confifting of twelve quills, and exceeding the wings eighteen lines. [A]

[A] Specific character of the Bulfinch, *Lóxia-Pyrrhula*:— " Its joints are black, the coverts of the tail and of the hind quills " of its wing white." Thus defcribed by Briffon: " Above ci- " nereous, below red, *(Male)*, wine-cinereous, *(Female)*: the " top of its head of a fhining black; its rump and lower-belly " white; its tail-quills violet-black, the lateral ones blackifh- " cinereous within, the outermoft on both fides marked interiorly " with a whitifh fpot." In England the Bulfinch breeds in the end of May or the beginning of June; it is pernicious to our gardens.

FOREIGN BIRDS,

WHICH ARE RELATED TO THE BULFINCH.

I.

The ORANGE GROSBEAK, *Lath.*

Le Bouveret, Buff.
Loxia Aurantia, Gmel.

I CLASS together two birds, the one from the Isle of Bourbon, and the other from the Cape of Good Hope: they resemble each other so closely, that they must belong to the same species; and there is, besides, an intercourse between these two places.

Black, and bright orange, are the prevailing colours in this bird, *fig.* 1. which I conceive to be the male: the orange is spread on the throat, the neck, and on all the body, without exception; black occupies the head, the tail, and the wings; but the wing-quills are bordered with orange, and some of them tipt with white.

In the female, all the head, the throat, and the fore-part of the neck, are covered with a ſort of black cowl; the under-part of the body is white; the upper-part orange, not ſo bright as in the male, but ſpreads diluting on the quills of the tail; the quills of the wings are delicately edged with light-gray, almoſt white: in both, the bill and legs are reddiſh.

Total length about four inches and a half; the bill ſomewhat leſs than four lines; the alar extent near ſeven inches; the tail twenty lines, conſiſting of twelve quills; it exceeds the wings about fifteen lines. [A]

[A] Specific character of the *Loxia Aurantia*:—"It is fulvous; "its cap, wings, and tail black."

II.

The WHITE-BILLED GROSBEAK, *Lath.*

Le Bouvreuil, à Bec Blanc, Buff.
Loxia Torrida, Gmel.

This is the only bird of Guiana that Sonini admits to be a true Bulfinch. Its bill, in the dried ſpecimen, is horn-colour; but we are aſſured that it is white in the living ſubject: the throat, the fore-part of the neck, and all the upper-part of the body, not excepting the wings and the tail, are black; on the wings is a ſmall white ſpot, which often lies concealed under the great coverts; the breaſt and belly are deep cheſnut.

This bird is of the ſize of our Bulfinch; its total length four inches and three-fourths, and its tail exceeds the wings by almoſt its whole length. [A]

[A] Specific character of the *Loxia Torrida*:—"It is black, "its breaſt and belly cheſnut."

III.

The LINEATED GROSBEAK, *Lath.*

Le Bouveron *, Buff.
Loxia Lineola, Linn. and Gmel.
Pyrrhula Africana Nigra minor, Briſſ.

This bird ſeems to form the ſhade between the European Bulfinches and the Round-bills of America. It is not larger than the Twite: a fine black, changing into green, is ſpread over the head, the throat, and all the upper-part of the body, including the coverts and quills of the tail and of the wings, or more properly ſpeaking on what appears of theſe; for the inſide is either not black, or is black of a different kind: there is alſo a very ſmall white ſpot on each wing, and three ſpots of the ſame colour, but larger, the one on the top of the head, and the

* Contracted for *Bouvreuil-Bec-rond,* (Bulfinch-Round-bill.)

two

two other below the eyes. All the under-part of the body is white; the feathers of the belly and the inferior coverts of the tail are frizzled in ſome ſubjects, for we may reckon the Frizzled Bulfinch of Brazil as belonging to this ſpecies, ſince the ſole difference conſiſts in the contexture of the plumage, which is ſuperficial and fluctuating. It is probable that Frizzled Bulfinches are cock-birds, Nature ſeeming in general to diſtinguiſh the males by ſtrength and beauty. But how, it would be aſked, is the male found in Brazil, and the female in Africa? I anſwer, 1. That the native climate of birds that paſs through different hands is very uncertain. 2. If theſe were brought alive to Paris, they might alſo have been carried from South America to Africa. Any perſon who will draw the compariſon between theſe two birds, will readily admit one of the two ſuppoſitions, rather than refer them to two different ſpecies.

Total length four inches and one-third; the bill four lines; the alar extent ſeven inches and a half; the tail twenty-one lines, conſiſting of twelve quills, and exceeding the wings about an inch. [A]

[A] Specific character of the *Loxia Lineola*:—"It is black; "the frontel line, and temples white."

IV. The

IV.

The MINUTE GROSBEAK, *Lath.*

Le Bec-Rond a Ventre Roux, Buff.
Loxia Minuta, Linn. and Gmel.
The Gray Loxia, Bancr. Guian.

America produces alſo Round-bills, which, though analogous to the Bulfinches, are ſtill ſo different as to merit a diſtinct denomination.

The preſent continues the whole year paired with its female. It is lively and tame, living near dwellings, and haunting grounds which have been under cultivation, but lately abandoned. It feeds upon grain and fruits; and, hopping about, it emits a cry much like that of the Sparrow, but ſhriller. It forms with a certain reddiſh herb a ſmall round neſt of about two inches diameter within, and places it in the ſame ſhrub that furniſhes its ſubſiſtence. The female lays three or four eggs.

The upper-part of the head, the neck, and the back, are brown-gray; the coverts of the wings, their quills, and thoſe of the tail, are of the ſame colour nearly, and bordered with white, or light cheſnut; the throat, the fore-part of the neck, the under-part of the body, the inferior coverts of the tail, and the rump, deep cheſnut; the bill and legs brown.

In ſome ſubjects the throat is of the ſame brown gray as the upper-part of the head. [A]

[A] Specific character of the *Loxia Minuta*:—"It is gray; its "rump, and the under-ſide of its body, ferruginous; the four, five, "or ſix wing-quills, white on both ſides at their baſe; its tail entire."

V. The

V.

The BLUE GROSBEAK.

Le Bec-Rond, ou *Bouvreuil-Bleu d'Amerique,* Buff.
Loxia Cærulea, Linn. and Gmel.
Pyrrhula Carolinensis Cærulea, Briss. and Klein.

Brisson mentions two American blue Bulfinches, of which he makes two separate species: but as they are both natives of the same climate, are of the same size, of the same shape nearly, and, except the wings, tail, and bill, are of the same blue colour; I shall consider them as forming one single species, and regard the differences as resulting from the influence of climate.

In both of them the prevailing colour is deep blue. The one from South America has a small black spot between the bill and the eye; the quills of the tail, those of the wings, and the great coverts of these, are edged with blue; the bill is blackish, and the legs gray.

That of North America has at the base of its bill a black circular zone, which extending forms a junction between the eyes; the quills of the tail and those of the wings, and their great coverts, brown, tinged with green; their middle coverts red, forming a transverse bar of the same colour; the bill brown, and the legs black. The plumage of the female is uniform, and of a deep brown, intermixed with a little blue.

With respect to the habits and œconomy of these birds, we can make no comparison, since we we are unacquainted with those of the first. Catesby informs us in regard to the one from Carolina, that it is very solitary and rare; that it continues paired with its female, and never appears in flocks; that it never winters in Carolina; that its song is monotonous, and consists of a repetition of the same note. In all these properties the analogy to our Bulfinch is marked. [A]

[A] Specific character of the *Loxia Cærulea*:—"It is cœrulean; "its wings dusky, with a purple bar at the base."

VI.

The BLACK GROSBEAK.

Le Bouvreuil ou *Bec-Rond Noire & Blanc*. Buff.
Loxia Nigra, Linn. Gmel. and Klein.
Pyrrhula Mexicana Nigra, Briss.
The Little Black Bulfinch, Catef. Alb. and Bancr.

As we have neither seen this bird, nor the dried specimen, we cannot decide whether it is a Bulfinch or a Round-bill. It has a little white on the anterior edge, and on the base of the two first quills of the wing; all the rest of the plumage is quite black, and even the bill and the legs; the upper mandible has a considerable scalloping on each side.

This bird is a native of Mexico. It is nearly of the size of the Canary Finch; total length five inches and one fourth; the bill five lines; the tail two inches, and exceeding the wings one inch. [A]

[A] Specific character of the *Loxia Nigra*:—"It is black, with "a white spot on the shoulder, and at the base of the two exterior "wing-quills."

VII. The

VII.

The PURPLE FINCH.

Le Bouvreuil ou *Bec-Rond Violet de la Caroline*, Buff.
Fringilla Purpurea, Gmel.
Pyrrhula Carolinensis Purpurea, Briss.

This bird is entirely of a dull violet, except the belly, which is white; the superior coverts of the wings where the violet is slightly mixed with brown, and the quills of the tail and of the wings which are parted by blue and brown, the former in the direction of their breadth, and the latter in that of their length.

The female is uniformly brown, only its breast is spotted, as in the Red-poll.

These birds appear in the end of November, and retire before the winter in small bodies. They live upon juniper-berries, and like our Bulfinches destroy the buds of the fruit-trees. They are nearly of the size of the Chaffinch.

Total length five inches and two-thirds; the bill five lines; the tail two inches, somewhat forked, consisting of twelve quills, and projecting seven or eight lines beyond the wings. [A]

[A] Specific character of the *Fringilla Purpurea*:—" It is olive: " its belly whitish; its wing-quills dusky within." It appears in Carolina in November, and feeds upon juniper-berries.

VIII.

The PURPLE GROSBEAK.

Le Bouvreuil ou *Bec-Rond Violet a Gorge & Sourcils Rouge*, Buff.
Loxia Violacea, Linn. and Gmel.
Pyrrhula Bahamensis Violacea, Briss.
Coccothraustes Purpurea, Klein.
Passer Niger Punctis Croceis, Ray, and Sloane.

This has still more of the violet than the preceding, for the quills of the wings and tail are also of that colour; but it is principally distinguished by its red throat, and its beautiful red eye-brows painted on the violet ground; the inferior coverts of its tail are also red, and its bill and legs are gray.

The female has the same red spots as the male, but the ground colour of its plumage is brown.

These birds are found in the Bahama Islands. They are nearly as large as a House-sparrow.

Total length five inches and two-thirds; the bill five or six lines; the tail two inches and a half, and projects thirteen or fourteen lines beyond the wings. [A]

[A] Specific character of the *Loxia Violacea*:—" It is violet; " its eye-brows, its throat, and its vent, white." It inhabits the Bahama Islands, and feeds upon the berries of the *Amyris Toxifera*, a tree from whose trunk a black poisonous juice exudes.

IX.

The BLACK-CRESTED GROSBEAK.

La Huppe Noire, Buff.
Loxia Coronata, Gmel.
Pyrrhula Americana Criſtata, Briſſ.

The plumage of this bird is painted with the richeſt colours; the head black, bearing a creſt of the ſame; the bill white; all the upper-part of the body brilliant red; the under-part fine blue; there is a black ſpot before the neck. This juſtifies the aſſertion of Seba, that it is inferior in beauty to none of the ſinging birds. We may thence infer that it has ſome ſort of warble. It is found in America.

Briſſon reckons it much larger than our Bulfinch. The meaſures were taken from a figure whoſe accuracy is not well aſcertained.—Total length ſix inches; the tail eighteen lines and more, and projecting about ſix lines beyond the wings. [A]

[A] Specific character of the *Loxia Coronata*:—" It is ſcarlet; " below cœrulean; the creſt on its head, and the middle ſpot on " its throat, black."

M

The

The HAMBURGH.

L'Hambouvreux, Buff.
Pyrrhula Hamburgensis, Briss.

Though this pretended Bulfinch is a native of Europe, I place it after thoſe of Africa and America, becauſe its habits are diſſimilar. It creeps upwards and downwards along the branches of trees like the Titmice; feeds upon horn-beetles, and other inſects; and has a tapered tail.

The upper-part of its head and neck is reddiſh brown, tinged with purple; its throat brown; it has a broad collar likewiſe brown upon a white ground; the breaſt is yellowiſh brown, ſprinkled with black longiſh ſpots; the belly, and the inferior coverts of the tail, white; the back, the ſhoulders, and all the upper-part of the body, like the breaſt; there are two white ſpots upon each wing; the quills of the wings are light brown and yellowiſh; thoſe of the tail, obſcure brown above, but white below; the iris yellow, and the bill black.

It is rather larger than the Houſe-ſparrow.—It is found near the city of Hamburgh.

Total length five inches and three-fourths; the bill ſix lines; the tail twenty-one lines, ſomewhat tapered, and extends almoſt its whole length beyond the wings.

N.º 105

THE COLY.

The COLY.

Le Coliou, Buff.

IT appears to us that this bird ſhould be ranged between the genus of the Widow and of the Bulfinch. Like the former, it has two long feathers in the middle of its tail; and the latter it reſembles by the form of its bill, which would be exactly that of the Bulfinch, were the lower mandible convex as the upper. But the tail of the Coly differs from that of the Widow, becauſe it conſiſts of tapered feathers, of which the two laſt project only three or four inches beyond the reſt; whereas the Widow-bird, beſides the true tail, which is a bundle of feathers of equal lengths attached to the rump, has appendices that in the different ſpecies of this genus contains two, four, and even ſix or eight feathers, extending to double or triple the length of the true tail. In the Widow-birds too the webs of the projecting feathers are equal on both ſides, and pretty long, and taper imperceptibly from the origin to the extremity, except in the Dominican and Shaft-tailed kinds; the former having its webs very ſhort, and quickly contracting towards the tips; the latter alſo very ſhort, but which uniformly lengthen and expand at the tips. In all the Colies, the feathers of the tail, whether thoſe which properly belong to it, or thoſe which project from it, have vanes that continually diminiſh from the inſertion to the termination. Thus the real affinity between the tail of the Widow-birds, and that of the Colies, conſiſts in its length; and this analogy is the cloſeſt in the caſe of the Dominican Widow.

M. Mauduit has made two important obſervations on this ſubject. The firſt is, that the long tails, the appendices, and other ornaments of certain birds, are not peculiar additions, but only the greater extenſion of the parts common to all the feathered tribes. Thus long tails conſiſt in the augmented growth of the quills of the tail; and creſts are derived from the enlargement of the feathers on the head. The muſtachoes of the Paradiſe bird appear to be only the production of the ſlender narrow plumules, which in all birds cover the *meatus auditorius externus**. The exuberant growth of the axillary feathers give riſe to the long floating feathers which ſhoot from under the wings in the Common Paradiſe, and of thoſe which preſent the double wings in the King Paradiſe. When theſe feathers lie flat, they ſtretch towards the tail; but when they are diſplayed they make angles with the axis of the body. They differ from all other feathers, as their webs are equal on both ſides of the ſhaft. They reſemble oars, and may ſerve to direct the bird's motion. And thus all the ornaments of plumage are derived from the mere exuberance or production of parts uſually leſs apparent.—

* *i. e.* The external aperture of the ear.

The

The ſecond obſervation of M. Mauduit is, that theſe decorations are rare in the cold and temperate climates in both hemiſpheres, but are pretty frequent in the hot countries, eſpecially in the old continent. Scarce any long-tailed birds are found in Europe, except the Pheaſant, the Cock, which alſo is often creſted, and has long floating feathers on the ſides; the Magpie, and the long-tailed Titmouſe; and we have hardly any creſted birds but the Eared Owls, the Lapwing, the Creſted Lark, and the Creſted Titmouſe. Some water-birds indeed, ſuch as the Ducks and Herons, have frequently long tails, or ornaments compoſed of plumes, tufts, and feathers floating on the rump.—Theſe are all the birds which the frigid and temperate zones afford, decorated with luxuriance of plumage. But in the torrid regions, eſpecially thoſe of the old continent, the greateſt number of birds are robed with ſuch rich attire: we might inſtance the Colies, the Paradiſe birds, the Kakatoes, the Widows, the Crowned Pigeons, the Hoopoes, the Peacocks, which are all natives of the hot climates of Aſia, &c.

The Colies belong to the ancient continent, and are found in the warm parts of Aſia and Africa, but never in America or Europe.

We are but imperfectly acquainted with four ſpecies or varieties, of which we can here give only the deſcriptions; for their habits and inſtincts are unknown.

1. *The Coly of the Cape of Good Hope* *, which we have described from a specimen in the king's cabinet. We cannot decide whether it is a male or a female. The body is entirely cinereous, pure on the back and rump, and mixed on the head; the throat and neck have a light lilac tint, which deepens on the breast; the belly is dirty white; the quills of the tail are cinereous, but the two lateral ones on each side are edged exteriorly with white; the two intermediate quills measure six inches and nine lines; those on the sides diminish gradually in length; the legs are gray, and the nails blackish; the bill is gray at its base, and blackish at its extremity. The length of the bird, including the long quills of the tail, is ten inches and three lines: so that the real size of the body exceeds not three inches and a half.—It is found at the Cape of Good Hope. [A]

[A] Specific character:—"The outmost wing-quills white externally; the body cinereous; below whitish."

2. *The Crested Coly of Senegal* † resembles the preceding, and may be regarded as a variety of that species, though it differs in point of size, being two inches longer. It has a sort of crest formed by projecting feathers on the head, and

* *Colius Capensis*, Gmel.
Loxia Capensis, Linn.
The Cape Coly, Lath.

† *Colius Senegalensis*. Gmel.
The Senegal Coly, Lath.

which

which is of the ſame complexion as the reſt of the body; there is a well-defined bar of fine ſky-blue behind the head, at the origin of the neck; the tail tapers from its baſe to its extremity; the bill is not entirely black; the upper mandible is white from its baſe to two-thirds of its length, and its end is black.—Theſe differences, though conſiderable, do not allow us to decide whether this bird be a different ſpecies, or only a variety of the preceding. [A]

[A] Specific character:—“ Mixed with pale gray and wine-“ colour; the tail bluiſh; the head creſted.”

3. Another ſpecies or variety rather larger than the laſt is the *Radiated Coly**, which we have ſeen in Mauduit's cabinet. It is thirteen inches long, including the long quills of the tail, which are only eight inches and a half, and exceed the wings by ſeven inches and a half; the bill is nine lines, black above, and whitiſh below.

It is called *Radiated Coly*, becauſe all the upper-part of the body is radiated firſt under the throat with brown bars on a ruſty gray ground, and under the belly with bars likewiſe brown on a rufous ground; the upper-part of the body is not radiated, is of a dirty gray, variegated ſlightly with lilac, which becomes more reddiſh on the rump and tail, which is green, and exactly ſimilar to that of the other Colies.

* *Colius Striatus*, Gmel.
The Radiated Coly, Lath.

Mauduit, to whom we owe our knowledge of this bird, thinks that it is a native of the countries near the Cape of Good Hope, becauſe it was brought from the Cape, with ſeveral other birds that certainly belong to that part of Africa. [A]

[A] Specific character:—"Gray; belly rufous, painted with "black tranſverſe ſtripes; tail green."

4. *The Coly of the Iſland of Panay**. We ſhall extract the account of it from Sonnerat's Voyage to New Guinea.

"It is of the ſize of the European Groſbeak; the head, the neck, the back, the wings, and the tail, are aſh-gray, with a yellow tinge; the breaſt is of the ſame colour, croſſed with blackiſh rays; the lower-part of the belly, and the upper-part of the neck are ruſty; the wings extend a little beyond the origin of the tail, which is extremely long, conſiſting of twelve unequal quills; the two firſt are very ſhort; the two following on each ſide are longer, and thus in the ſucceſſive pairs till the two laſt, which exceed all the reſt; the fourth and fifth pairs differ little from each other, in regard to length; the bill is black; the legs are pale fleſh colour; the feathers that cover the head are narrow, and pretty long, and form a creſt, which the bird can raiſe or depreſs at pleaſure." [B]

[B] Specific character:—"Cinereous, tinged with yellow; be-"low rufous; breaſt ſtriped with black; head creſted."

* *Colius Panayenſis*, Gmel.
The Panayan Coly, Lath.

N.º 106

FIG. 1. THE MANAKIN. FIG. 2, THE CRESTED MANAKIN.

The MANAKINS.

Pipræ, Linn.

THESE birds are ſmall and handſome; the largeſt are not equal in ſize to a Sparrow, and the others are inferior to that of the Wren. The general characters are theſe: the bill is ſhort, ſtrait, and compreſſed on the ſides near the tip; the upper-mandible is convex above, and ſlightly ſcallopped on the edges, rather longer than the lower-mandible, which is plain and ſtraight.

In all theſe birds, the tail is ſhort and ſquare-cut, and the toes have the ſame diſpoſition as the Cock of the Rock, the Tody, and the Calao, viz. the mid-toe is cloſely connected to the outer-toe by a membrane, as far as the third joint, and the inner-toe as far as the firſt joint only. But as much as in that circumſtance they reſemble the Cock of the Rock, ſo much are they removed from the Cotingas: yet ſome authors have ranged the Manakins with the Cotingas *; others have joined them with the Sparrows †, with the Titmice ‡, with the Linnets §, with the Tanagres ||, and with the Wren ¶; other

* Edwards. † Klein. ‡ Linnæus. § Klein.

|| Marcgrave, Willughby, Johnſton, Salerne, &c.

¶ Gerini.

other nomenclators are more culpable for denominating them *Pipra*, or for claſſing them together with the Cock of the Rock*, to which they bear no analogy, except in this diſpoſition of the toes and in the ſquare ſhape of the tail: for, beſides the total diſproportion in ſize, the Cock of the Rock being as large, compared with the Manakins, as the common hen contraſted with a Sparrow, there are many other obvious characters which diſtinguiſh them: their bill is much ſhorter in proportion, they are generally not creſted, and in thoſe which have a creſt, it is not double, as in the Cock of the Rock, but formed by ſingle feathers, ſomewhat longer than the reſt. We ought therefore to remove from the Manakins, not only the Horn-bills, but the Cock of the Rock, and reckon them an independent genus.

The natural habits common to them all were not known, and the obſervations which have been made are ſtill inſufficient to admit an exact detail. We ſhall only relate the remarks communicated to us by Sonini of Manoncour, who ſaw many of theſe birds in their native climates. They inhabit the immenſe foreſts in the warm parts of America, and never emerge from their receſſes to viſit the cleared grounds or the vicinity of the plantations. They fly with conſiderable ſwiftneſs, but always at a ſmall height,

* Briſſon.

and to ſhort diſtances; they never perch on the ſummits of trees, but on the middle branches; they feed upon ſmall wild fruits, and alſo eat inſects. They generally occur in ſmall bodies of eight or ten of the ſame ſpecies, and ſometimes intermingled with other flocks of the ſame genus, or even of a different genus, ſuch as the Cayenne Warblers, &c. It is commonly in the morning that they are found thus aſſembled, and then ſeem to be joyous, and warble their delicate little notes; the freſhneſs of the air ſeems to inſpire the ſong, for they are ſilent during the burning heat of the day, and diſperſe and retire to the ſhade of the thickeſt parts of the foreſt. This habit is obſerved indeed in many kind of birds, and even in thoſe of the woods of France, where they collect to ſing in the morning and evening; but the Manakins never aſſemble in the evening, and continue together only from ſun-riſe to nine or ten o'clock in the forenoon, and remain ſeparate during the reſt of the day and the ſucceeding night. In general, they prefer a cool humid ſituation, though they never frequent marſhes or the margins of lakes.

The name of the *Manakin* was beſtowed on theſe birds by the Dutch ſettlers at Surinam. We know ſix diſtinct ſpecies, but we can only give the firſt the appellation which it has in its native region.

The TIGE, or GREAT MANAKIN.

Le Tijé, ou *Grand Manakin*, Buff.
Pipra-Pareola, Linn. Gmel. and Browsk.
Manacus Cristatus Niger, Briss.
Tijé-guacu of Marcgrave, Will.
The Blue-backed Manakin, Lath. and Edw.

First Species.

This species has been well described by Marcgrave. It is the largest of all the Manakins; its total length is four inches and a half, and it is nearly of the bulk of a Sparrow; the upper-part of its head is covered with fine red feathers, longer than the rest, and which the bird can erect at pleasure, which gives it the appearance of being crested; the back and the small superior coverts of the wings are of a beautiful blue, the rest of the plumage is velvet black; the iris is of a fine sapphire colour; the bill is black, and the legs are red.

The Abbe Aubry, Rector of St. Louis, has in his cabinet a bird by the name of *Tise-guacu of Cuba*, which is perhaps a variety of the present, arising from the difference of age or sex; the only distinction being, that the large feathers on the upper-part of the head are of a diluter red, and even somewhat yellowish. The designation given to it would seem to imply that it is found in the Iſland of Cuba, and perhaps in other parts of America; but it is very rare at Cayenne, and is a bird of ſhort flight, and therefore it could hardly paſs from the continent to Cuba.

The Green Manakin with a red creſt is the young of this ſpecies; ſeveral Manakins have been obſerved, whoſe plumage was interſperſed with blue feathers, but the green is obſcure. Theſe birds muſt be frequent in the warm climates of America, for we often received them along with other birds. [A]

[A] Specific character of the Blue-backed Manakin, or *Pipra-Pareola*: — "Its creſt is blood-colour, its body black, its back "blue."

The NUT-CRACKER.

Le Caſſe-Noiſette, Buff.
Pipra-Manacus, Linn. and Gmel.
Manacus, Briſſ.
The Black-capped Manakin, Edw. and Lath.

Second Species.

We apply this name, becauſe the cry of this bird is exactly like the noiſe made by the ſmall inſtrument with which we crack nuts. It has no other ſong or warble; it is pretty common in Guiana, eſpecially in the ſkirts of the foreſts; for, like the other Manakins, it ſhuns the ſavannas

vannas and cleared grounds. The Nut-crackers live in ſmall flocks with the other Manakins, but intermingle not with them; they remain for the moſt part on the ground, and rarely perch on the branches, and then they uniformly prefer the low ones. They ſeem to live more upon inſects than fruits: they are often found among the lines of ants, which ſting their feet, and make them hop and utter their nut-cracking cry, which they repeat very often. They are very lively and friſky; they are ſeldom at reſt, though they only hop, and cannot fly far.

The plumage is black on the head, the back, the wings, the tail, and white on all the reſt of the body; the bill is black and the legs yellow. [A]

[A] Specific character of the Black-caped Manakin, *Pipra-Manacus*: — "Black, below white, ſpot on the neck and the "wings white."

The RED MANAKIN.

Le Manakin Rouge, Buff.
Pipra-Aureola, Linn. and Gmel.
Manacus Ruber, Briſſ.
The Red and Black Manakin, Edw. and Lath.

Third Species.

The male is of a fine vivid red on the head, the neck, the upper-part of the back, and the

the breaſt; orange on the forehead, the ſides of the head, and the throat; black on the belly, with ſome red and orange feathers on the ſame part; black alſo on the reſt of the upper-part of the body, the wings, and the tail; all the quills of the wings, except the firſt, have near the middle, and on the inſide, a white ſpot, which forms a bar of the ſame colour when the wing is diſplayed; the top of the wings is of a very deep yellow; their inferior coverts are yellowiſh; the bill and legs are blackiſh.

In the female, the upper-part of the body is olive, with a ſlight appearance of a red crown on the head; the under-part of the body is olive-yellow: the ſhape and bulk are the ſame as in the male.

In the young bird, all the body is olive, with red ſpots on the forehead, the head, the throat, the breaſt, and the belly.

It is the moſt common of all the ſpecies of Manakins in Guiana. [B]

[B] Specific character of the Red Manakin, *Pipra Aureola*: "Black, head and breaſt ſcarlet, white ſpot on the fore-part of "the wing-quills."

The ORANGE MANAKIN.

Le Manakin Orangé, Buff.
Pipra-Aureola, var. Linn.
Manacus Aurantius, Briſſ.
The Black and Yellow Manakin, Edw. and Lath.

Fourth Species.

Edwards is the firſt author who has given a figure of this bird; but he was miſtaken in ſuppoſing it to be the female of the preceding. We have juſt deſcribed the female of the red Manakin, and the preſent is undoubtedly a different ſpecies, ſince it is extremely rare in Guiana, whereas the Red Manakin is very common. Linnæus has fallen into the ſame error, becauſe he only copies Edwards.

The head, the neck, the throat, the breaſt, and the belly are of a fine orange, all the reſt of the plumage being black; only the wings are marked with ſome white ſpots as in the Red Manakin; like that bird too, it has blackiſh legs, but a white bill, ſo that notwithſtanding the ſimilarity in the bar on the wings, in the colour of the legs, and in the bulk and ſhape of the body, it cannot be regarded as a variety of the Red Manakin.

I. The

I. The GOLD-HEADED MANAKIN *.

II. The RED-HEADED MANAKIN †.

III. The WHITE-HEADED MANAKIN ‡.

Fifth Species.

We preſume that theſe three birds are only varieties of the ſame ſpecies, for they are exactly of the ſame ſize, being only three inches and eight lines in length; whereas all the preceding ſpecies, which have been placed in the order of their magnitude, are four inches and a half, and four inches and three-fourths, &c. Theſe three are likewiſe of the ſame ſhape, and even their colours are ſimilar, if we except thoſe of the head, which in the firſt are of a fine yellow, in the ſecond of a bright red, and in the third of a beautiful blue; there is no other ſenſible difference in the plumage, which is all uniformly of a fine gloſſy black: the feathers which cover the thighs are of a pale yellow, with an oblong ſpot of bright red on the exterior ſurface. In the

* This is the *Pipra Erythrocephala* of Linnæus and Gmelin, the *Manacus Aurocapillus* of Briſſon, and the *Gold-headed Black Titmouſe* of Edwards.

† This is a variety of the *Pipra Erythrocephala* of Linnæus, and Willughby's *ſecond kind of Tangara*.

‡ This is the *Pipra Leucocilla* of Linnæus, in his laſt edition, the *Parus Pipra* in the eleventh edition, the *Pipra Leucocapilla* of Gmelin, and the *White-capped Manakin* of Edwards and Latham.

the firſt indeed, the bill is whitiſh and the legs black; in the ſecond, the bill is black and the legs cinereous; and in the third, the bill is brown-gray, and the legs reddiſh: but theſe ſlight differences do not appear ſufficient to conſtitute three ſeparate ſpecies, and perhaps one of them is only the female of another. However, Mauduit, to whom I communicated this article, aſſured me that he never ſaw, in the White-headed Manakin, the red feathers that cover the knee in the Golden-headed Manakin: if this difference were invariable, we might infer that theſe formed two ſeparate ſpecies of Manakins: but Sonini aſſured us, that he has ſeen White-headed Manakins with red feathers on their knees, and there is ſome reaſon to ſuſpect that the ſpecimens obſerved by Mauduit were mutilated.

Theſe Manakins are found in the ſame ſituations, and are pretty common in Guiana. It would even appear that the ſpecies is ſpread through the ſeveral other warm countries, as Brazil and Mexico. We have learnt nothing particular in regard to their inſtincts and œconomy. We are certain only that, like the other Manakins, they conſtantly live in the woods, and that they have the chirping common to the whole genus, except the Nut-cracker. [A]

[A] Specific character of the *Pipra Gutturalis*: — "It is "black, its throat white."

The

The WHITE-THROATED MANAKIN.

Pipra Gutturalis, Linn. and Gmel.
Manacus Gutture Albo, Briss.

VARIETY.

This differs from the preceding by the colours of its head, which is glossy black like the rest of the plumage, except a kind of white collar which rises on the neck, and tapers to a point on the breast. It is exactly of the same size as the three preceding, being only three inches eight lines in length. We know not from what country it is brought, having seen it only in private cabinets, where it was mentioned by this name without any other indication. Sonini never met with it in Guiana; however, it is probably, like the three others, a native of the warm climates in America.

The VARIEGATED MANAKIN.

Manacus Serena, Gmel.
Manacus Alba Fronte, Briss.
The White-fronted Manakin, Lath.

Sixth Species.

We have given the epithet of *variegated*, because the plumage is interspersed with spots of different colours, which are all very beautiful and diftinct. The forehead is of a fine dead white; the crown of the head is of a beryl colour; the rump of a brilliant blue; the belly of a fhining orange, and all the reft of the plumage of a fine velvet black; the bill and legs are black. It is the pretticft and fmalleft of all the Manakins, being not more than three inches and a half long, and not bigger than a Wren. It is found in Guiana, whence we received it; but it is very rare, and we are unacquainted with its habits. [A]

[A] Specific character of the *Manacus Serena*:—"It is black, "its front white, its rump fea-green, its belly fulvous."

Befides the fix fpecies and their varieties, which we have now defcribed, modern nomenclators apply the name of Manakin to four birds mentioned by Seba, and which we fhall here notice, only to fhew the errors into which fuch artificial claffifications lead.

The firft is thus defcribed by Seba:—

*Bird called Miacatototl by the Brazilians**.

"Its body is ornamented with blackifh feathers, and its wings with feathers of turkey-blue; its head is blood-coloured, and has a golden-yellow collar round the neck and throat; the

* This is the *Pipra Torquata* of Gmelin, the *Manacus Torquatus* of Briffon, and the *Collared Manakin* of Latham.

bill

bill and legs are of a pale yellow." Brisson, who had not seen the bird, adds the dimensions and other details, which are furnished neither by Seba, nor by any other author. It is also strange that Seba should bestow the appellation of *Miacatototl* upon this bird, which he says was brought from Brazil; for that word is not in the language of that country, but is a Mexican word, and signifies *the Maize-bird*. But that Seba was mistaken in this application is evinced by the circumstance that Fernandez employs the same term in describing a Mexican bird which is very different. His words are these:—

Of the Miacatototl, or *Maize-bird*.

"It is pretty small; so called because it usually sits upon the shoots of maize; the belly is palish, and the rest of its body black, but with white feathers interspersed; its wings and tail are ash-coloured below. It frequents cool places, and is good eating." Seba has manifestly confounded therefore two different birds under the same name. But the description of Fernandez is as imperfect as that of Seba, whose figure is still worse; so that it is impossible to decide the arrangement.

The same is the case with another bird mentioned by Seba, under the name of

Rubetra, or *Crested American Bird* *.

* This is the *Pipra Rubetra* of Linnæus and Gmelin, the *Manacus Cristatus Rufus* of Brisson, and the *Yellow Manakin* of Latham.

"It is not one of the smallest singing birds," says he; "it has a yellow crest, its bill too is yellow, except the under-mandible, which is brown, the plumage round the neck, and on the body, is of a yellow rufous; the tail, and the great quills of the wings, are of a shining blue, while the small quills are pale-yellow." From this description of Seba, Brisson has ventured to range this bird among the Manakins; but if he had inspected the figure, though it is a very bad one, he would have perceived the tail to be very long, the bill thin, curved, and elongated; characters quite different from those of the Manakins. I should therefore conclude, that this bird is still farther removed from the Manakins than the preceding.

A third bird which our nomenclators have reckoned a Manakin, is that mentioned by Seba under the name of

Picicitli, or *The Least Crested Bird of Brazil* *.

"The body and wings are purple, which here and there is deeper; the crest is a most beautiful yellow, and formed by a small tuft of feathers; its bill is pointed, and its tail red. In short, this little bird is very handsome, whatever view we take of it." From this confused description, Brisson concludes the bird to be a Manakin, and adds the dimensions and other circum-

* This is the *Pipra Cristata* of Linnæus and Gmelin, the *Manacus Cristatus Ruber* of Brisson, and the *Purple Manakin* of Latham.

circumſtances without citing his authorities; yet Seba tells us that its bill is pointed; and his figure is very imperfect. Beſides, he is miſtaken in aſſerting that it is a Brazilian bird, for the name *Picicitli* is Mexican; and Fernandez employs the ſame word to deſcribe another bird which is certainly Mexican.

"The *Picicitli* is likewiſe a native of Tetzcoqua; it is very ſmall, and its body wholly cinereous, except its head and neck, which are black, with white ſpots encircling the eyes (which are large), the front extending on the breaſt. Theſe birds appear after rain. If they be raiſed in the houſe they ſoon die. They have no ſong. They are excellent food; but the Indians are uncertain where they breed."

It is eaſy to ſee that there is no relation between this bird and that of Seba, who has very injudiciouſly occaſioned a confuſion of terms.

The ſame remark may be made with regard to the fourth bird deſcribed by Seba, under the name of

Coquantototl, or *Small Creſted Bird, ſhaped like a Sparrow* *.

"Its bill is yellow, ſhort, curved, and bent backwards. There is a yellow ſpot above the eyes; the ſtomach and belly are of a pale bluiſh yellow; the wings are of the ſame colour, and

* This is the *Pipra Griſea* of Gmelin, the *Manacus Criſtatus Griſeus* of Briſſon, and the *Gray Manakin* of Latham.

 mixed

mixed with some slender carnation feathers, but the principal feathers are ash-gray; the rest of the body is gray; there is a small crest behind the head." From this indication Brisson has inferred that the bird is a Manakin; but the shape of the bill is alone sufficient to evince the contrary; and besides, since it is shaped like a Sparrow, its form must be different from that of the Manakins. It is evident, therefore, that this bird, whose name also is Mexican, is widely removed from the genus of the Manakins.—We shall not venture at present to assign the rank of the four birds, but wait till inquisitive travellers may have thrown light upon the subject.

SPECIES RELATED TO THE MANAKIN.

The WHITE PLUME.

Le Plumet Blanc.
Pipra Albifrons, var. Linn. and Gmel.
The White-faced Manakin, var. Lath.

THIS ſpecies is new. It is found in Guiana, but rare. M. de Manoncour preſented a ſpecimen to the king's cabinet. It is diſtinguiſhed by a very long white creſt, conſiſting of feathers about an inch in length, and which it erects at pleaſure. It differs from the Manakins by its ſize, being ſix inches long; whereas the largeſt of the Manakins is only four inches and a half; the tail too is long and tapered, which in the Manakins is ſhort and ſquare; and the bill is much longer in proportion, and more hooked than that of the Manakins. Indeed, the only property in which it reſembles the Manakins is, the arrangement of the toes; and but for this character it might be ranged with the Ant eaters: we may regard it as forming the intermediate ſhade. We are unacquainted with its œconomy.

The CINEREOUS BIRD OF GUIANA.

Pipra Atricapilla, Gmel.
The Black-crowned Manakin, Lath.

This ſpecies is new. We ſhall only obſerve, that it ought not to be regarded as a true Manakin; for its tail is much longer, and tapered, and its bill is longer; but as it reſembles the Manakins in the diſpoſition of its toes, and in the figure of the bill, we ought to place it after them.

This bird is found in Guiana, but is not frequent. It was preſented by M. De Manoncour to the King's cabinet.

The PAPUAN MANAKIN, *Lath.*

*Le Manikor**, Buff.
Pipra Papuenſis, Gmel.

This is a new ſpecies brought from New Guinea to the King's cabinet by Sonnerat. It differs from the Manakins, as the two middle quills of the tail are ſhorter than the lateral ones, and as it wants the ſcallop that occurs on the upper-mandible in all the Manakins: ſo that we ought to exclude it from the genus of the Manakins, eſpecially as theſe birds, which are natives of America, are not probably found in New Guinea.

* This name is contracted for *Manakin Orangé,* Buffon having taken it at firſt for a Manakin.

The

The upper-part of the body is black, with greeniſh reflections; the under-part of the body is of a dirty white; there is an oblong orange ſpot on the breaſt, which extends as far as the belly; the bill and legs are black. But Sonnerat could give no information with reſpect to its manner of living.

The COCK OF THE ROCK.

Le Coq de Roche, Buff.
Pipra-Rupicola, Linn. and Gmel.
Rupicola, Briss.
Upupa Americana Lutea, Ger. Orn.
Felsenhahn, Walch. Natur.
The Hoopoe Hen, Edw.
The Crested Manakin, Penn.
The Rock Manakin. Lath.

THOUGH this bird is of an uniform colour, it is one of the most beautiful of South America; for this colour is very beautiful, and its plumage nicely tapered. It feeds upon fruits, perhaps for want of seeds; since it differs from the gallinaceous tribe by the shape of its toes only, which are connected by a membrane, the first and second as far as the third joint, and the second with the third no farther than the first joint. The tail is very short and square, as are some coverts of the wings; several of these feathers have a sort of fringe on each side, and the first great feather of each wing is scalloped from the tip to one third of its length: but what characterizes it the best is, a beautiful crest on the crown, longitudinal, and of a semicircular form. In the minute descriptions which Brisson and Vosmaër have given, this crest is imperfectly depicted; it is not single, but double, consisting of

No. 107

THE ROCK MANAKIN.

of two inclined planes that meet at the vertex. In other reſpects their deſcriptions are accurate, except that they are applicable only to the male. The plumage of the male is of a fine red; that of the female is entirely brown, only there are ſome ſhades of ruſt-colour on the rump, the tail, and the feathers of the wings. The creſt of the female is alſo double, but not ſo thick, ſo tall, ſo round, or ſo far protruded on the bill, as in the male. Both male and female are generally thicker and larger than the Ring-pigeon; but the different individuals probably vary in ſize; for Briſſon repreſents this bird of the bulk of a Roman-pigeon, and Voſmaër aſſerts that it is rather ſmaller than the Common Pigeon. This difference might alſo be occaſioned by the manner of ſtuffing the ſpecimens; but in the natural ſtate the female, though ſomewhat ſmaller than the male, is undoubtedly much larger than the Common Pigeon.

The male aſſumes not the fine red till he has attained ſome age; the firſt year he is only brown like the female; but as he grows up, his plumage becomes marked with points and ſpots of rufous, which gradually deepen into red, though perhaps perfected before advanced age.

Though this bird muſt have attracted the notice of all who ever ſaw it, no traveller has deſcribed its habits and œconomy. Sonini de Manoncour is the firſt who has obſerved it with attention. It lodges not only in the deep clefts

of

of the rocks, but even in the large dark caverns which totally exclude the ſolar rays; a circumſtance that has induced many to believe it to be a nocturnal bird; whereas it flies and ſees diſtinctly in the day-time: however, it ſeems naturally to prefer gloomy receſſes, ſince it is moſt frequent in caves which cannot be entered without the light of torches. We may therefore preſume, that their eyes are conſtructed like thoſe of cats, and adapted both for the day and the night. The male and female are equally lively, and extremely wild. It is impoſſible to ſhoot them, unleſs the perſon be concealed behind a rock, where he muſt often wait ſeveral hours before an opportunity occurs; for the inſtant they perceive him, they eſcape to a diſtance by a flight which is rapid, though rather low and ſhort. They feed upon ſmall wild fruits, and uſually ſcrape the ground, clap their wings, and ſhake themſelves like the dunghill fowls; but they neither crow like the cock, nor cluck like the hen. Their cry reſembles the ſound of the ſyllable *ké*, pronounced with a ſhrill drawling tone. They conſtruct their neſt rudely in the holes of rocks with ſmall dry ſticks; and commonly lay two white ſpherical eggs, which are of the ſize of thoſe of pigeons.

The males emerge oftener from their caverns than the females, which ſeldom appear, and probably do not quit their retreats except in the night. They can eaſily be tamed, and M. de Manoncour

Manoncour ſaw one at the Dutch-poſt on the river Maroni, which was allowed freely to live and run about with the poultry.

They are found in great numbers on the mountain Luca, near Oyapoc, and on the mountain Courouaye, near the river Aprouack; and theſe are the only parts in this region of America whence we can expect to procure theſe birds. They are much eſteemed for the ſake of their beautiful plumage, and are very ſcarce and dear; becauſe the ſavages and negroes, whether from ſuperſtition or fear, will not venture into the dark caverns where they lodge. [A]

[A] Specific character of the *Pipra Rupicola*:—"It has an erect "creſt, with a purple margin; its body is ſaffron; the coverts of "its tail are truncated."

The PERUVIAN COCK OF THE ROCK.

Pipra Peruviana, Lath.

There is another ſpecies, or rather variety of the Cock of the Rock, which is found in the provinces of Peru: its tail is much longer than that of the preceding, and its feathers have not ſquare ends; its wings are not fringed; inſtead of an uniform red, both wings and tail are black, and the rump cinereous; the creſt alſo is different, being lower and conſiſting of detached feathers: in other reſpects, this Peruvian bird reſembles the Guiana Cock of the Rock ſo cloſely, that

that we may regard it as a variety of the ſame ſpecies.

We might conſider theſe birds as the repreſentatives of our dunghill poultry in the New World; but I am told that, in the interior parts of Guiana and Mexico, there are wild fowls which bear ſtill more analogy. Theſe are indeed much ſmaller, being ſcarcely of the ſize of a Common Pigeon; they are generally brown and rufous; but they have the ſame ſhape, the ſame little fleſhy comb on the head, and the ſame port as our ordinary fowl; their tail is alſo ſimilar in ſhape and poſition, and the males have the crow of the Cock, though feebler. The ſavages who inhabit the remote tracts are perfectly well acquainted with theſe birds, but have never reduced them to the domeſtic ſtate; nor is this in the leaſt ſurpriſing, for they have tamed none of the animals which might have proved uſeful, eſpecially the Hoccos or Curaſſos, the Marails and the Agamis, among the birds; and the Tapirs, the Pecaris, and the Agamis, among the quadrupeds. On the contrary, the ancient Mexicans, who were civilized, domeſticated ſome animals, and particularly theſe ſmall brown fowls. Gemelli Carreri relates, that they were called *Chiacchialacca;* and he ſubjoins, that they were preciſely like our common poultry, only rather ſmaller, and their feathers browniſh.

The COTINGAS.

FEW birds have ſuch beautiful plumage as the Cotingas; all thoſe who have had an opportunity of ſeeing them, whether travellers or naturaliſts, ſeem to have been charmed, and ſpeak of them with rapture. Nature has ſelected her choiceſt and her richeſt colours, and ſpread them with elegance and profuſion: the painting glows with all the tints of blue, of violet, of red, of orange, of purple, of ſnow-white, and gloſſy black; ſometimes theſe tints melt into each other by the ſweeteſt gradations; at other times they are contraſted with wonderful taſte: the various reflexions heighten and enliven the whole. The merit is intrinſic; it is expreſſive; it is inimitable.

All the ſpecies, or, if we chuſe, all the branches of the brilliant family of the Cotingas belong to the New Continent; and there is no foundation for what ſome have alleged, that they are found in Senegal. They appear to delight in warm countries; they ſeldom occur ſouth of Brazil, or roam north of Mexico: and conſequently they would hardly traverſe the immenſe ſtretch of ocean that ſeparates the continents in thoſe latitudes.

All that we know of their habits is, that they never perform diſtant journies, but have only periodical flittings, which are confined within a narrow circle: they appear twice a-year in the plantations; and though they arrive nearly at the ſame time, they are never obſerved in flocks. They generally haunt the ſides of creeks in ſwampy ground *, which has occaſioned ſome to call them water-fowls. They find among the aquatic plants abundance of inſects, on which they feed, and particularly what are termed *karias* in America, and which, according to ſome, are wood-lice; and according to others, a ſort of ants. The creoles have, it is ſaid, more motives than one for hunting after theſe birds;—the beauty of the plumage, which pleaſes the eye; and, according to ſome, the delicacy of the fleſh, which flatters the palate. But it is difficult to obtain both; for the plumage is often ſpoiled in attempting to ſkin the bird; and this is probably the reaſon why ſo many imperfect ſpecimens are now brought from America. It is ſaid that they alight among the rice-crops and do conſiderable injury: if this be true, the creoles have ſtill another reaſon for deſtroying them †.

* Mr. Edwards, who was unacquainted with the œconomy of the Cotingas, conjectured, from the ſtructure of their feet, that they frequented marſhes.

† The little which I have related of the habits of the Cotingas was communicated by M. Aublet: but I muſt add, that M. de Manoncour heard that the fleſh of the Cotingas was much eſteemed at Cayenne; perhaps this is true only of ſome ſpecies.

The

No. 108

THE PURPLE-BREASTED CHATTERER.

The ſize varies in the different ſpecies, from that of a ſmall Pigeon to that of a Red-wing, or even under; in all of them the bill is broad at the baſe; the edges of the upper-mandible, and often thoſe of the lower, are ſcalloped near the tip; the firſt *phalanx* of the outer-toe joined to that of the mid-toe; and, laſtly, in moſt of them, the tail is a little forked or notched, and conſiſts of twelve quills.

M

The BLUE RIBAND.

Le Cordon Bleu, Buff.
Ampelis-Cotinga, Linn. Gmel. and Briſſ.
The Purple-breaſted Manakin *, Penn. Edw. and Lath.

A bright blue is ſpread on the upper-part of the body, of the head, and of the neck, on the rump, the ſuperior coverts of the tail, and the ſmall coverts of the wings; the ſame colour appears alſo on the inferior coverts of the tail, the lower-belly, and the thighs. A fine violet purple covers the throat, the neck, the breaſt, and a part of the belly, as far as the thighs; and on this ground is traced, at the breaſt, a belt of the ſame blue with that of the back, and which has procured this bird the appellation of *Blue Riband*, or

* Called alſo the *Thruſh of Rio-Janeiro*, and the Creoles term it *Hen of the Woods*.

or *Knight of the Holy Ghost*. Below the first belt there is in some subjects another of a beautiful red, besides many flame-spots on the neck and the belly: these spots are not disposed regularly, but scattered with that negligence in which nature seems to delight, and which art labours in vain to imitate.

All the quills of the tail and of the wings are black, but those of the tail, and the middle ones of the wings, are edged exteriorly with blue.

The specimen which I observed was brought from Brazil; its total length was eight inches; its bill ten lines; its alar extent thirteen inches; its tail two inches and two-thirds, composed of twelve quills, and projecting eighteen lines beyond the wings. The one described by Brisson was every way somewhat larger, and of the size of a thrush.

The female has neither of these belts; nor has it the flame-spots on the belly and breast *. In every other respect it resembles the male; the bill and legs of both are black, and the ground-colour of their feathers is blackish, and that of the purple feathers white; and the *tarsus* is covered behind with a sort of down. [A]

* "At Cayenne there are two other (Blue-Riband Thrushes)," says Salerne, "which resemble this exactly, except that the one "wants these spots, and the other the Blue-Riband."

[A] Specific character of the *Ampelis-Cotinga*:—"It is of a very "bright blue, below purple, its wings and tail black."

M

The

The PURPLE THROATED CHATTERER.

Le Quereiva, Buff.
Ampelis Cayana, Linn. and Gmel.
Cotinga Cayanenſis, Briſſ.
Lanius Ococolin, Klein and Seba.

The greateſt portion of each of its feathers, reckoning from their inſertion, is black; but as the tips are beryl, this is really the predominant colour of the plumage. In ſome parts of the upper ſurface of the body the dark hue ſtrikes through the coverts, but then it forms only ſmall ſpecks; and it is totally concealed by the blue in the under-ſurface of the body: only, in ſome ſpecimens, there are, near the rump and the thighs, a few ſmall feathers, which are partly black, and partly purple-red.

The throat and a part of the neck are covered with a broad ſpot of a very bright violet-purple, which in different ſubjects varies in extent. The coverts of the wings, their quills, and thoſe of the tail, are almoſt all black, edged or tipt with beryl; the bill and legs are black.

This bird is found in Cayenne; it is of the ſize of the Red-wing, and faſhioned like the preceding, except that the wings, when cloſed, reach not the middle of the tail, which is rather longer. [A]

[A] Specific character of the *Ampelis Cayana*:—" It is bright " blue; its neck violet below."

The BLUE-BREASTED CHATTERER.

La Terſine, Buff.
Ampelis Terſa, Linn. and Gmel.

Linnæus is the firſt, and even the only one, who has hitherto deſcribed this bird: the head, the top of the neck, the quills of the wings and of the tail, are black; the throat, the breaſt, the lower-part of the back, the outer edge of the quills of the wings, light blue: there is a tranſverſe bar of light blue on the ſuperior coverts of the ſame quills; the belly is yellowiſh-white, and the ſides are of a deeper caſt. Linnæus does not inform us from what country it is brought; but it is probable that it is a native of America, like the other *Cotingas*. I ſhould be even tempted to regard it as a variety of the preceding, ſince blue and black are the prevailing colours of the upper-part of the body, and the colours of the under-part are dilute, as uſual in the females, the young birds, &c. A ſight of the ſubject would be neceſſary to decide the queſtion. [A]

[A] Specific character of the *Ampelis Terſa*:—"It is bright "blue, its back black, its belly yellowiſh-white."

The SILKY CHATTERER.

Le Cotinga à Plumes Soyeuſes, Buff.
Ampelis Maynana, Linn. and Gmel.
Cotinga Mayanenſis, Briſſ.

Almoſt all the feathers in the body of this bird, and the coverts of the wings and of the tail are unwebbed, and parted into filaments; ſo that they reſemble ſilky briſtles more than real feathers: a property which is ſufficient to diſtinguiſh it from all the other Cotingas. The general colour of its plumage is bright blue, varying into a fine ſky-blue, as in the preceding; but we muſt except the throat, which is deep violet, and the quills of the tail and of the wings, which are blackiſh; moſt of theſe are edged exteriorly with blue; the feathers of the head and of the upper-part of the neck are long and narrow, and the ground-colour is brown; that of the feathers of the body and breaſt, &c. conſiſts of two colours; at the inſertion of theſe feathers it is white, and then purple-violet, which in ſome parts ſtrikes through the blue of the incumbent feathers; the bill is blue, and the legs are black.

Total length ſeven inches and one-third; the bill nine or ten lines; the tarſus the ſame; the alar extent thirteen inches and one-third; the tail about three inches, conſiſting of twelve quills, and exceeds the wings by an inch. [A]

[A] Specific character of the *Ampelis Maynana*:—" It is bright " blue, its throat violet."

M The

The POMPADOUR CHATTERER.

Le Pacapac, ou *Pompadour*, Buff.
Ampelis-Pompadora, Linn. Gmel. and Borowſk.
Cotinga Purpurea, Briſſ.
Turdus Puniceus, Pall.

All the plumage of this beautiful bird is bright gloſſy-purple, except the quills of its wings, which are whitiſh tipt with brown; and the inferior coverts of the wings, which are entirely white: the under-ſide of the tail is of a lighter purple; the ground of the feathers on every part of the body is white; the legs are blackiſh; the bill gray-brown, and on each ſide of its baſe riſes a ſmall whitiſh ſtreak, which, paſſing under the eyes, bounds the face.

The great coverts of the wings are oddly faſhioned, long, narrow, ſtiff, pointed, and ſpout-ſhaped; their vanes parted, their ſhaft white, and without webs at its tip, which reſembles in ſome degree the appendices that terminate the wing in the Common Chatterer (*Jaſeur*), and is nothing but the projection of the ſhaft beyond the webs. This is not the only point of reſemblance between theſe two ſpecies; in the ſhape of their bill, their ſize, the proportional dimenſions of their tail, their feet, &c; but their inſtincts are very different, ſince the common Chatterer prefers the mountains, and all the ſpecies of *Cotingas* frequent the low marſhy grounds.

Total

Total length ſeven inches and a half; the bill ten or eleven lines; the tarſus nine or ten lines; the alar extent above fourteen inches; the tail two inches and a half, conſiſting of twelve quills, and projecting frôm ſix to eight lines beyond the wings.

The Pompadour is migratory; it appears in Guiana near the inhabited ſpots in March and September, when the fruits on which they feed are ripe; they lodge among the large trees on the banks of rivers, and neſtle on the higheſt branches, but never retire into the wide foreſts.—The ſpecimen from which this deſcription was made came from Cayenne. [A]

[A] Specific character of the *Ampelis Pompadora*: — "It is "purple; the neareſt coverts of its wings are ſword-ſhaped, elon-"gated, boat-ſhaped, and ſtiff."

M

VARIETIES *of the* POMPADOUR.

Pacapac Gris-Pourpre, Buff.

I. The Grey-Purple Pompadour. It is rather ſmaller than the preceding, but its proportions are exactly the ſame; the great coverts of its wings have the ſame ſingular conformation, and it inhabits the ſame country. So many common properties leave no room to doubt, that, though the plumage be different,

 theſe

these two birds belong to the same species; and since the present is smaller, I should be apt to suppose it to be a young one that has not acquired its full growth, or the finished colours of its plumage: all that was purple in the preceding is, in the present, variegated with purple and cinereous; the under-side of the tail is rose-coloured; the quills of the tail are brown: what appears of those of the wings are also brown; the interior and concealed part of their shaft is white from its insertion to two-thirds of its length; and also the middle ones are edged exteriorly with white.

II. M. Daubenton the younger and myself have seen, at Mauduit's, a Gray Cotinga, which appeared to belong to the species of the Pompadour, and to be only younger than the preceding, but which ought not to be confounded with another which is also called the Gray Cotinga, and which I shall presently describe under the name of *Guirarou* *.

It is probable that these are not the only varieties which exist of this species, and that others will be found among the females of different ages.

* M. de Manoncour has verified our conjectures on the spot. In his last voyage to Cayenne, he found that the Purple-gray Cotinga is the young bird, and that it takes at least eighteen months to acquire its full colour.

The

The RED CHATTERER.

L'Ouette, ou *Cotinga Rouge de Cayenne*, Buff.
Ampelis-Carnifex, Gmel.
Lanius Ruber Surinamensis, Ger.
Icterus Totus Ruber, Klein.
Cotinga Rubra, Briss.
Red Bird from Surinam, Edw.

The prevailing colour of its plumage is red, but diversified by various tints, which it assumes in different parts; the most vivid, which is scarlet, is spread over the upper-part of the head, and forms a sort of crown or cap, of which the feathers are pretty long, and are conjectured by Edwards to rise like a crest: the same scarlet covers the lower-part of the belly, the thighs, the lower-part of the back, and almost to the end of the tail-quills, which are tipt with black; the sides of the head, the neck, the back, and the wings are shaded with-deep tints, which change the red into a fine soft crimson; but the darkest cast is a sort of border which surrounds the scarlet cap, and this is a little more dilute behind the neck and on the back, and more so on the throat and breast; the coverts of the wings are edged with brown, and the great quills become more and more obscure, and terminate almost in black; the bill is a dull red; the legs dirty yellow; and, what is remarkable, the tarsus is covered with a sort of down as far as the origin of the toes.

The

The Red Cotinga migrates, or rather flits, like the Pompadour, only it is more common in the interior parts of Guiana.

Total length about ſeven inches; the bill nine lines; the legs ſeven lines; the tail two inches and a half, and projects twenty lines beyond the wings, and conſequently the alar extent is leſs than in the preceding ſpecies. [A]

[A] Specific character of the *Ampelis Carnifex*: — "It is red; "the ſtripe at its eyes, and the tips of the quills of the wings "and of the tail, are black."

M

The CARUNCULATED CHATTERER,

Le Guira Panga, ou *Côtinga Blanc*, Buff.
Ampelis Carunculata, Gmel.
Cotinga Alba, Briſſ.

Laët is the only perſon who has mentioned this bird, and all that he ſays amounts to no more than that its plumage is white and its cry very loud. Since his time, the ſpecies has been in a manner loſt, even in Cayenne; and M. de Manoncour has the merit of re-diſcovering it.

Both the male and the female are figured in the *Planches Enluminées*. They were perched upon trees beſide a ſwamp when they were killed; they were betrayed by their cry, which,

as

as Läet * obſerved, was very loud; and it reſembled the ſound of the two ſyllables *in an*, uttered with an exceedingly drawling tone.

The moſt remarkable character of theſe birds is, a ſort of caruncle under the bill as in the turkies, but differently organized: it is flaccid and pendulous when the bird is compoſed at reſt; but when the paſſions are rouzed, it ſwells in every dimenſion, and, in this ſtate of tenſion, is more than two inches long, and three or four lines in circumference at the baſe: this effect is produced by air, which is driven through an aperture of the palate into the cavity of the caruncle and inflates it.

This caruncle differs from that of the turkey alſo in another circumſtance; it is covered with ſmall white feathers; and beſides, it is not peculiar to the male. The plumage of the female is however entirely different: for in the male the bill and legs are black, all the reſt of a pure ſpotleſs white, except ſome tints of yellow on the rump, and on ſome of the quills of the tail and of the wings: but in the female the colour is not ſo uniform; the upper-part of the head and body, the ſuperior coverts of the wings, and moſt of the quills of the wings and of the tail, are olive mixed with gray; the la-

* Voyagers ſay, that its voice reſembles the ſound of a bell, and may be heard at the diſtance of half-a-league. *Hiſt. Gen. des Voyages*, tom. xiv. p. 299.

teral

teral quills of the tail gray, edged with yellow; the cheeks and forehead white; the feathers of the throat gray, edged with olive; thofe of the breaft and of the anterior part of the belly gray, edged with olive and tipt with yellow, and the coverts of the lower furface of the tail lemon-yellow; the inferior coverts of the wings white, edged with the fame yellow.

The male and female are nearly of the fame fize. Total length twelve inches; length of the bill eighteen lines; its breadth at its bafe feven lines: length of the tail three inches nine lines, confifting of twelve equal quills, and projecting twenty-one lines beyond the wings. [A]

[A] Specific character of the *Ampelis Carunculata*:— "It has "a pendulous, expanfible, and moveable caruncle at the bafe of the "bill."

M

The VARIEGATED CHATTERER.

L'Averano, Buff.
Ampelis Variegata, Gmel.
Cotinga Nævia, Briff.
Guira-Punga, Ray and Will.

The head is deep brown; the quills of the wings blackifh; their fmall coverts black; the great coverts blackifh, with fome mixture of brownifh green: all the reft of the plumage is cinereous, mixed with blackifh, chiefly on the back, and with greenifh on the rump and tail.

The

The bill is broad at the baſe, as in the Cotingas; its tongue is ſhort; its noſtrils uncovered; its iris bluiſh-black; its bill black; its legs blackiſh. It has ſeveral black fleſhy appendices under the neck, nearly of a lance-ſhape, which marks a ſlight affinity to the preceding at the ſame time that it diſcriminates it from all the other Cotingas.

The Variegated Cotinga is as large as a Pigeon; the length of its bill, which is an inch, equals the greateſt breadth; its legs are twelve or thirteen lines; its tail is three inches, and is almoſt wholly beyond the reach of the wings.

The female is rather ſmaller than the male, and has not the fleſhy appendices under the neck; it reſembles the Fieldfare in ſhape and ſize; its plumage is a mixture of blackiſh, of brown, and light green; but theſe colours are diſtributed ſo, that the brown predominates on the back, and the light-green on the throat, the breaſt, and the under-part of the body.

Theſe birds grow plump and juicy. The male has a very ſtrong voice, and inflected in two different ways; ſometimes it reſembles the noiſe occaſioned by ſtriking a cutting inſtrument againſt a wedge of iron (*kock*, *kick*); and ſometimes it is like the jarring of a bell that is cracked (*kur*, *kur*, *kur*). It is heard in no part of the year but during the ſix weeks of the middle of ſummer; that is, in the ſouthern hemiſphere, in December and January; and hence the Portu-

guese

guefe name, *Ave de Verano*, i. e. *Bird of Summer*. It is remarked that its breaft is marked exteriorly with a furrow which runs through its whole length ; and alfo that its wind-pipe is very wide, which perhaps contributes to the ftrength of its voice. [A]

[A] Specific character of the *Ampelis Variegata* :—" It is cine-" reous ; to its throat are attached two lance-fhaped caruncles."

M

The GUIRAROU, *Buff.*

Lanius-Nengeta, Linn. and Gmel.
Cotinga Cinerea, Briſſ.
The Gray Pye of Brazil, Edw.
The Gray Shrike, Penn. and Lath.

If the beauty of plumage formed the characteriſtic feature of the Cotingas, this bird, and that of the preceding article, would be regarded as degenerate branches of the original ſtock. The *Guirarou* has nothing remarkable either in its colours, or in their diſtribution, if we except a black bar below the eyes, and the tint of the iris, which is ſapphire: a uniform light gray is ſpread over the head, the neck, the breaſt, and all the under-part of the body; the thighs, and the upper-part of the body, cinereous; the quills, and coverts of the wings, blackiſh; the quills of the tail black, tipt with white, and its ſuperior coverts white; laſtly, the bill and legs are black.

The flat ſhape, and the ſhortneſs of the bill, the loudneſs of its voice, which is ſomewhat like that of the Blackbird, but ſhriller, and its haunting the margin of water, are the chief circumſtances in which the *Guirarou* reſembles the Cotingas; its ſize is alſo nearly the ſame, and it inhabits the ſame climates: yet Willughby has referred it to the White-ears; and other excellent ornithologiſts have reckoned it a Fly-catcher.

For my own part, I ſhall not venture to aſſign its genus; I ſhall retain the name which it bears in its native climate, and wait for fuller obſervations made on ſeveral living ſubjects, which will point out its proper arrangement. The *Guirarous* are very common in the interior parts of Guiana, but are not found at all in Cayenne; they ramble little; many occur generally in the ſame diſtrict; they perch generally on the loweſt branches of certain large trees, where they pick up ſeeds and inſects, on which they ſubſiſt. From time to time, they cry all at once, allowing an interval between each ſound; this cry, though harſh in itſelf, is cheering muſic to travellers who have loſt their way in the immenſe foreſts of Guiana, for it directs them to the banks of a river.

The ſubject obſerved by M. De Manoncour was nine inches and a half total length; its bill twelve inches long, ſeven broad, five thick at the baſe, and encirled with hairs; the tail was ſquare, four inches long, and exceeded the wings by two inches and a half; the *tarſus* was an inch, and ſo was the bill *

* I owe theſe details to M. de Manoncour.

VARIETY

VARIETY of the GUIRAROU.

I know of one only; it is what we have called the *Gray Cotinga*; and Daubenton and myſelf ſuſpect that it is a variety of age, becauſe it is ſmaller, its total length being ſeven inches and a half, and its tail rather ſhorter, the wings reaching to the middle, and all the other differences reſult from defect. It has neither the black bar under the eyes, nor the white-bordered tail, nor the white ſuperior coverts; the quills of the wings are edged with white, but they are not ſo blackiſh; and thoſe of the tail not ſo black as in the *Guirarou*. [A]

[A] Specific character of the *Lanius-Nengeta*: — "Its tail "wedge-ſhaped, with a white tip; its body is cinereous; below, "white."

M

The ANTERS.

Les Fourmiliers, Buff.

IN the low, ſwampy, thin-ſettled lands of South America, the ſwarms of inſects and loathſome reptiles ſeem to predominate over all the reſt of the animal creation. In Guiana and Brazil * the ants are ſo aſtoniſhingly multiplied, that their hills are ſome fathoms wide, and ſeveral feet in height, and proportionally populous as thoſe of Europe, of which the largeſt are only two or three feet in diameter; ſo that they may be computed to contain two or three hundred times the number of ants. Yet they exceed ſtill more in number; and in the wilderneſſes of Guiana they are an hundred times more frequent than in any part of the ancient continent.

* This is alſo the caſe in many other parts of America. Piſo relates, that in Brazil, and even in the wet grounds of Peru, the quantity of ants is ſo enormous, that they devour all the ſeeds which are committed to the earth; and though fire and water be employed to extirpate them, the attempts have hitherto failed of ſucceſs. He adds, that it were much to be wiſhed that Nature had ordained in thoſe countries many ſpecies of animals like the Ant-eaters (*Myrmecophagæ*, Linn.), which might bore into the hillocks, and extract theſe inſects with their long tongue. Some of the ants are not larger than thoſe of Europe; others are twice or thrice as large. They raiſe hills as large as hay-ſtacks; and their number is ſo vaſt, that they make tracks ſeveral feet broad in the fields, and in the woods, and often through an extent of many leagues.—Fernandez ſays alſo that theſe ants are larger, and pretty much like our winged-ants, and that their hills are of an incredible height and width.

But

But (ſuch is the ſyſtem of Nature !) every creature is the deſtined prey of another; and generation and deſtruction are ever conjoined. We have in the former work given an account of the *Tamanoir*, of the *Tamandua*, and of the other quadrupeds which feed upon ants; we are now to write the hiſtory of a kind of birds which live alſo upon theſe inſects.—We were unacquainted with the exiſtence of the Anters till M. de Manoncour preſented the ſpecimens to the King's cabinet.

The Anters are natives of Guiana, and are analogous to none of the European birds; but in the ſhape of their body, of their bill, of their feet, and of their tails, they bear a great reſemblance to the ſhort-tailed Thruſhes *(Breves)*, which our nomenclators have improperly confounded with the Blackbirds: but as the ſhort-tailed Thruſhes inhabit the Philippines, the Moluccas, the iſland of Ceylon, Bengal, and Madagaſcar, it is more than probable that they are not of the ſame race with the Anters of America. Theſe appear indeed to conſtitute a new genus, for which we are wholly indebted to M. de Manoncour, whom I have ſo often cited for his extenſive knowledge of foreign birds: he has preſented above an hundred and ſixty different ſpecies to the Royal cabinet; and has alſo been ſo obliging as to communicate to me all the obſervations which he made in his voyages to Senegal and America. I have on many

 occa-

occaſions availed myſelf of this information; and in particular I have formed entirely from it the hiſtory of the Anters.

In French Guiana, and indeed in all countries where natural hiſtory is little known, names are applied to animals from the ſlighteſt analogies. This has been the caſe with the Anters: they were obſerved to perch ſeldom, and run like Partridges; but as they were inferior to theſe birds in ſize, they were diſtinguiſhed at Cayenne by the appellation of *Little Partridges.*

But theſe birds are neither Partridges, nor Blackbirds, nor ſhort-tailed Thruſhes; only they reſemble the laſt in their chief external characters. Their legs are long; their tail and wings ſhort; the nail of the hind-toe more hooked, and longer than thoſe of the fore-toes; the bill ſtrait and lengthened; the upper-mandible ſcalloped at its extremity, which bends at the junction of the lower mandible, and projects about a line beyond it; but their tongue is ſhort, and beſet at the tip with ſmall cartilaginous and fleſhy threads. Their colours are alſo very different; and it is very probable that their inſtincts are diſſimilar, ſince they inhabit widely diſtant climates. When we deſcribed the ſhort-tailed Thruſhes, we were unable to give any account of their natural habits, ſince no travellers had taken notice of them, and therefore we cannot draw any compariſon with thoſe of the American Anters.

In

In general the Anters keep in flocks, and feed upon ſmall inſects, and chiefly ants, which are for the moſt part ſimiliar to thoſe of Europe. They are almoſt always found upon the ant-hills, which in the interior tracts of Guiana, are more than twenty feet in diameter, and whoſe inſect nations retard the extenſion of cultivation, and even conſume the proviſions of life.

There are ſeveral ſpecies of Anters, which, though very different in appearance, often aſſociate together; the large ones and the ſmall, the long-tailed and the ſhort-tailed, are found on the ſame ſpot. Indeed, if we except the principal kind, which are very few, it is rare to find in the reſt two ſubjects perfectly alike; and we may ſuppoſe that this diverſity ariſes from the intermixture of the ſmall ones: ſo that we muſt regard them as mere varieties, and not diſtinct ſpecies.

In all theſe birds the wings and tail are very ſhort, and therefore ill calculated for flying; accordingly they only trip along the ground, and hop among the low branches; and though lively and active, they never ſhoot through the air.

The voice of the Anters is various in the different ſpecies, and in ſome it is very ſingular.

As inſects are the chief food of theſe birds, they ſeek the ſolitary tracts where thoſe are not moleſted by the intruſion of man, and ſwarm in abundance. They live in the thickeſt and the

 remoteſt

remoteſt foreſts, and never viſit the ſavannas, the cleared grounds, and ſtill leſs the neighbourhood of plantations. They employ dry herbs careleſsly interwoven in the conſtruction of their neſts, which are hemiſpherical, and two, three, or four inches in diameter, and ſuſpend them by the two ſides on the buſhes, two or three feet from the ground. They lay three or four eggs, which are almoſt round.

The fleſh of moſt of theſe birds is unpalatable food, and has an oily rank taſte, and when opened, the digeſted maſs of ants, and of other inſects they ſwallow, exhales a putrid offenſive ſmell.

The KING OF THE ANTERS.

Le Roi des Fourmiliers, Buff.
Turdus Rex, Gmel.
Turdus Grallarius, Lath. Ind.
The King Thruſh, Lath. Syn.

Firſt Species.

This is the largeſt and the moſt unfrequent of all the birds of this genus. It is never ſeen in flocks, and ſeldom in pairs; and as it is generally alone among the others, and is larger than them, it is called *The King of the Anters.* It is the more entitled to that appellation, as it affects an uncommon diſtance to other birds, and even to thoſe of its own kind. If ſo excellent an obſerver as M. de Manoncour had not communicated the details of its manner of living, to diſcover

cover it to be an Anter, from the mere inſpection, would have been almoſt impoſſible; for its bill is thicker, and differently ſhaped from that of all the others. This bird is generally on the ground, and is far from being ſo lively as the reſt, who hop around it. It frequents the ſame ſpots, and feeds alſo upon inſects, eſpecially ants. The female, as in all the other ſpecies of this genus, is larger than the male.

Its length from the point of the bill to the end of the tail is ſeven inches and a half; its bill is brown, ſomewhat hooked, fourteen lines long, and five lines thick at the baſe, which is beſet with ſmall whiſkers; the wings extend the whole length of the tail, which is only fourteen lines; the legs are brown, and two inches long.

The under-ſide of the body is variegated with brown rufous, blackiſh, and white; the rufous brown is the predominant colour as far as the belly, where it grows dilute, and the whitiſh prevails. Two white bars deſcend from the corners of the bill along with the duſky ſhade of the throat and neck; on the breaſt is a white ſpot nearly triangular. The upper-ſide of the body is brown rufous, ſhaded with black and white, except on the rump and tail, where the colour is uniform.—The ſize and the tints are ſubject to vary in different ſpecimens, and we have only deſcribed here the more uſual appearances. [A]

[A] Specific character of the *Turdus Rex*:—" Its plumage conſiſts of brown and rufous; below more dilute; the back of its head lead-coloured; its front variegated with white and brown."

The AZURIN.

Turdus Cyanurus, Gmel.
The Blue-tailed Thrush, Lath.

Second Species.

We have described this bird after the Blackbirds, and have nothing to add to the former account. We remarked that it was undoubtedly not a Blackbird; and from its external appearance it ought to be ranged among the Anters. We are unacquainted with its œconomy. It is rare in Guiana, but was sent however from thence to M. Mauduit.

The GREAT BELFRY.

Le Grand Béfroi, Buff.
Turdus Tinniens, Gmel.
The Alarum Thrush, Lath.

Third Species.

We apply the epithet of *Great* only to distinguish it from another smaller species; for its total length exceeds not six inches and a half; its tail is sixteen lines, and projects six lines beyond the wings; its bill is eleven lines, black above and white below, and three lines and a half broad at the base; the legs are eighteen lines

No. 109

THE ALARUM THRUSH.

lines long, and, as well as the toes, are of a light lead-colour.

The tints vary in almoſt each individual, and the dimenſions are alſo variable *;—we have ſtated the average.

In this ſpecies the females are much larger than the males, and ſtill more diſproportioned than in the firſt ſpecies: in this reſpect the Anters reſemble the birds of prey.

What moſt remarkably diſtinguiſhes this bird, which we have named *Belfry*, is, the ſingular ſound that it makes in the evenings and mornings: this reſembles the din of an alarum-bell. Its voice is ſo ſtrong, that it can be heard at a great diſtance, and one would hardly ſuppoſe it emitted by ſo ſmall a bird. The ſucceſſion of ſounds, which is as rapid as the quick ſtrokes of a bell, continues about an hour. It appears to be a ſort of call ſimilar to that of the Partridges, only it is heard at all ſeaſons, and every day, at the riſing of the ſun, and before his ſetting: however, as the period of love is not fixed in thoſe hot climates, the Partridges, as well as the Anters, have their call in every ſeaſon indiſcriminately.

The King of the Anters and the Belfry are the only birds of the genus that are palatable food. [A]

* In ſome individuals, the upper-mandible, though ſcalloped and a little hooked, exceeds not the under.

[A] Specific character of the *Turdus Tinniens*:—" It is brown " above, white below, its breaſt ſpotted with black, its tail " equal."

The

The SMALL BELFRY.

Le Petit Béfroi, Buff.
Turdus Lineatus, Gmel.
The Speckled Thrush, Lath.

VARIETY.

Its length five inches and a half; the upper-part of the body is olive, which grows more dilute on the rump; the tail, of which the quills are brown, as well as those of the wings, exceeds these by ten lines; the under-part of the throat is white, and the feathers below become gray, and spotted with rusty brown as far as the belly, which is entirely rusty brown.

From this description it is easy to perceive the striking resemblance of colours between this bird and the Great Belfry, and the figure is precisely the same.

The PALIKOUR, *or* ANTER, *properly ſo called.*

Turdus Formicivorus, Gmel.
The Ant Thruſh, Lath.

Fourth Species.

It is near ſix inches long; its body not ſo thick as that of the Little Belfry, and its bill longer than in that ſpecies; its iris reddiſh, and its eyes encircled by a ſkin of ſky-blue; the legs and the lower mandible of the ſame colour.

The throat, the fore-part of the neck, and the top of the breaſt, are covered with a cravat of black, with a black and white border, which extends behind the neck and forms a half-collar; the reſt of the under-part of the body is cinereous.

The birds of this ſpecies are very lively, but fly not more than the others in open air; they climb among the buſhes like magpies, expanding the feathers of their tail.

They make a ſort of quavering, interrupted by a feeble cry, which is abrupt and ſhrill.

Their eggs are brown, and nearly as large as thoſe of ſparrows; the great end is ſprinkled with ſpots of a deep brown; the neſt is thicker and cloſer interwoven than thoſe of the other Anters, and is covered externally with more than one layer of moſs.

The

The COLMA, *Buff.*

Turdus Colma, Gmel.
The Rufous-naped Thrush, Lath.

The Colma may likewiſe be conſidered as a variety of the preceding, or as a cloſely-related ſpecies: all the plumage of its body is brown; below it is brown-gray, and on the belly cinereous; only on the lower-part of the head, behind the neck, there is a ſort of rufous half-collar, and the throat is white, dotted with brown gray. We have formed its name *Colma*, from this laſt character. In ſome ſubjects the rufous half-collar is wanting.

The TE'TE'MA, *Buff.*

Turdus Colma, Variety.

This is a native of Cayenne, and ſeems to reſemble much the preceding, not only in ſize, which is the ſame, and in ſhape, which is nearly ſimilar, but in the diſpoſition of the colours, which are almoſt the ſame on all the upper-part of the body. The greateſt difference occurs on the throat, the breaſt, and the belly, which are blackiſh brown: whereas in the Colma, the origin of the neck and throat are white, variegated

gated with ſmall brown ſpots, and the breaſt and belly are aſh-gray, which would induce us to ſuppoſe that the differences are only ſexual. In that caſe I ſhould reckon the Tétéma as the male, and the Colma as the female, becauſe its colours are generally more dilute.

The CRESTED ANTER.

Le Fourmilier Huppé, Buff.
Turdus Cirrhatus, Gmel.
The Black-creſted Thruſh, Lath.

Fifth Species.

The average length of this bird is near ſix inches: the upper-part of the head is decorated with long black feathers, which it can erect at pleaſure like a creſt; the iris is black, the under-part of the throat is covered with black and white feathers; the breaſt and the under-part of the neck are black;—all the reſt of the body is aſh-gray.

The tail is two inches four lines in length, and conſiſts of twelve tapered quills *, edged and tipt with white, and exceeds the wings an inch, whoſe ſuperior coverts are tipt with white, and, in ſome ſubjects, they are of the general colour of the body, or aſh-gray.

* In all the ſpecies of the Anters, the tail is more or leſs tapered; thoſe which have it larger than the reſt, have it alſo thinner, and the quills weaker.

 The

The female has alſo a creſt, or rather the ſame long feathers on the head, but they are rufous, and its plumage differs from that of the male in nothing except a ſlight ſhade of ruſty upon the gray.

Theſe birds have a cluck like that of a pullet; they lay three eggs *, and breed ſeveral times annually.

* M. de Manoncour found in the month of December ſeveral young of this ſpecies ready to fly. He tried in vain to rear ſome of them; for they all died in the ſpace of four days, though they ate very heartily crumbs of bread.

The WHITE-EARED ANTER.

Turdus Auritus, Gmel.
Pipra Leucotis, Gmel.
The White-eared Manakin, Lath.
The White-eared Thruſh, Lath.

Sixth Species.

It is four inches nine lines in length; the upper-part of the head is brown, and the lower ſides of the fore-part of the head and throat are black: a ſmall bar of ſhining white ſtretches from the poſterior angle of the eye to below the head, where the feathers are broader and longer than thoſe of the head.

There is nothing remarkable in the reſt of the plumage: the colour of the upper-part of the body is an unpleaſant mixture of olive and ruſty.

rufty. The fuperior part of the under-fides of the body is rufous, and the reft gray.

The tail is fifteen lines in length; the wings extend its whole length; the legs are brown: the habits of the bird are the fame as thofe of the preceding kinds.

The CHIMER.

La Carrillouneur, Buff.
Turdus Tintinnabullatus, Gmel.
Turdus Campanella, Lath. Ind.
The Chiming Thrufh, Lath. Syn.

Seventh Species.

The total length of this bird is four inches and a half, and its tail projects nine lines beyond the wings.

Befides the habits common to the Anters, the Chimer has others peculiar to itfelf: it haunts the grounds where the ants abound, but does not intermingle with the reft; it generally forms fmall feparate parties of four or fix: they hop about and utter a very fingular cry, exactly like the chime of three different-toned bells: their voice is fonorous, confidering the fmallnefs of their fize. We might fuppofe that they fing their parts, though it is likely that each founds fucceffively the three notes; but we are not certain, as no perfon has ever been at the trouble to domefticate mesticate them. Their voice is not so loud as that of the Great Belfry, which is indeed equal to that of a bell of considerable size; and the Chimers are not distinctly audible farther than fifty paces, while the Belfry may be heard at the distance of half a league. These birds continue their chiming without intermission for whole hours.

The species is very rare, and found only in the still forests, in the heart of Guiana. [A]

[A] Specific character of the Chimer, *Turdus Tintinnabulatus*: —"Its crown and temples white, spotted with black, its eye-"brows black, its chin white, its breast carnation, spotted with "black; its back, its wings, and its tail, brown; its :ump, its "belly, and vent, orange-rufous."

The BAMBLA.

Turdus-Bambla, Gmel.
The Black-winged Thrush, Lath.

Eighth Species.

We have given it this name, because there is a white transverse bar on each wing (*bande-blanche*). The habits of the bird are unknown; but from its resemblance to the other Anters, I should infer that it belongs to the same genus, though still a distinct species.

Besides these eight species of Anters, we have seen three others which were brought from Cayenne,

Cayenne, but without the leaſt account of their natural habits. [A]

[A] Specific character of the *Turdus Bambla*:—" It is ſpotted, " above it is duſky-rufous, below cinereous, its wings black, and " has a tranſverſe white ſtripe."

The ARADA, *Buff.*

Turdus Cantans, Gmel.
The Muſician Thruſh, Lath.

This was called by M. de Manoncour, the *Muſician of Cayenne*; I rather chuſe to retain the name of *Arada*, which it receives in its native country.

It is not exactly an Anter; but we have placed it after theſe, becauſe it has the ſame external characters, though it differs in its habits. It perches upon trees, and never alights on the ground, except to pick up ants and other inſects, upon which it feeds. It is diſtinguiſhed from them by a remarkable property; for all the Anters utter harſh cries without any moleſtation, while the Arada has the moſt charming warble. It commences often with the ſeven notes of the octave, and then whiſtles different ſoft varied airs, which are lower than thoſe of the Nightingale, and more like the breathing of a ſweet toned flute; and it is ſaid to excel even that celebrated choriſter of the grove in delicate tender

 melody.

melody. It has alſo a ſort of whiſtle, reſembling that by which a perſon calls upon another: travellers frequently miſtake this ſound, and, by following it, they are led more aſtray; for as they approach, the bird continually recedes, and whiſtles at intervals.

The Arada avoids ſettled ſpots; it lives alone in the depth of the vaſt foreſts, and the ſoftneſs of its melody ſeems in ſome meaſure to relieve the gloomy ſtillneſs around. It is one of the very few birds in the New World which Nature has diſtinguiſhed by the charms of its ſong. But the ſpecies is not numerous; and the traveller may frequently purſue his pathleſs journey without meeting a ſingle Arada to ſooth his ſympathetic gloom.

The colours of its plumage correſpond not with the richneſs of its ſong; they are dull and obſcure.—The total length is four inches, and the tail is radiated tranſverſely with rufous, brown, and blackiſh;—it exceeds the wings by ſeven lines.

To the Arada we may refer a bird which Mauduit ſhewed to us: it reſembles that in the length and ſhape of its bill, the form of its tail, the length of its legs, in having ſome white feathers mixed with the brown ones on the ſides of the neck; the ſize is nearly equal, and the ſhape ſimilar; but the tip of its bill is more hooked, its throat is white, with a half-collar of black below, and its plumage is uniform, and

and not ſtriped with brown lines, as in the Arada, whoſe throat and under-part of its neck are red. We may preſume therefore, that this bird is either a diſtinct variety of the Arada, or a contiguous ſpecies, ſince it inhabits Cayenne; though, being unacquainted with its habits, we ſhall not at preſent preſume to decide the matter. [A].

[A] Specific character of the *Turdus Cantans*: — " It is " brown-rufous, variegated with blackiſh tranſverſe ſtreaks, be- " low partly white; its chin, its cheeks, and its throat, orange- " rufous; a black ſpace ſpotted with white on each ſide of the " neck."

The NIGHTINGALE ANTERS.

Les Fourmiliers Roſſignols, Buff.

In their external figure theſe birds are intermediate between the Anters and the Nightingales: their bill and feet are like thoſe of the Anters, and their long tail reſembles that of the Nightingale's. They live in flocks in the vaſt foreſts of Guiana; they run upon the ground and hop among the low branches, but fly not in open air; they feed upon ants and other ſmall inſects; they are very nimble, and when they friſk about, they make a ſort of quavering, ſucceeded by a feeble ſhrill cry, which they repeat ſeveral times when they call upon each other.

We know only two ſpecies.

The CORAYA, *Buff.*

Turdus Corava, Gmel.
The Barred-tail Thrush, Lath.

First Species.

We have given it this name, becaufe its tail is radiated tranfverfely with blackifh *(queue-rayée)*. The length is five inches and a half, from the point of the bill to the end of the tail; the throat and the fore-part of the neck are white; the breaft is lefs white, and receives a cinereous fhade; there is a little of rufty under the belly and on the thighs; the head is black and the upper-part of the body rufous-brown; the tail is tapered, and two inches long, and extends at leaft eighteen lines beyond the wings; the hind nail is, as in the Anters, the longeft and ftrongeft of all.

The ALAPI, *Buff.*

Turdus Alapi, Gmel.
The Black-headed Thrush, Lath.

Second Species.

It is rather larger than the preceding, being fix inches long: its throat, the fore-part of its neck and breaft are black; the reft of the under-part of the body cinereous; an olive-brown is fpread ſpread over the upper-part of the head, neck, and back; the reſt of the upper-part of the body is deeper cinereous than that of the belly: there is a white ſpot on the middle of the back; the tail is blackiſh and ſomewhat tapered, projecting one inch and a half beyond the wings, the quills of which are brown above and below, and the ſuperior coverts are of a very deep brown, dotted with white, whence its name *Alapi* (*ailes Piquetées*).

The female has not the white ſpot on the back; its throat is white, and the reſt of the under-part of the body ruſty, with aſh-gray feathers on the ſides of the lower-belly, and on what form the inferior coverts of the tail; the points of the coverts of the wings are alſo ruſty, and that of the upper-part of the body is not ſo deep as in the male.

Theſe ſhades, and even the colours themſelves, vary in different ſubjects, as we have had occaſion to obſerve with regard to the Anters.

The AGAMI*, *Buff.*

Psophia Crepitans, Linn. Gmel. and Borowsk.
Grus Psophia, Pall.
Phasianus Antillarum, Briss.
L'Oiseau Trompette, Descr. Surin.
The Gold breasted Trumpeter, Lath.

To avoid confusion, we shall restore to this bird the name of *Agami*, which it has ever received in its native region. In a preceding part of the present work, we were deceived by the account of Father Dutertre, and have mentioned it by the appellation of *Caracara;* but that term was bestowed by Marcgrave upon a bird of prey totally different from the *Agami.*

Naturalists have entertained the most opposite opinions with regard to this bird. Dutertre supposes it to be a Pheasant; Barrere reckons it a Wild Hen; Pallas terms it a Crane; and Adanson seems to insinuate that it is a large aquatic bird of the genus of the Lapwing, because its knees are prominent, and its hind-toe is placed a little higher than the three fore-toes, and because it appears the intermediate kind between the *Jacana* and the *Kamichi.*

* It is called *Trompetero* by the Spaniards of the province of Maynas, and *Agami* by the French at Cayenne.

But

N.° 110

THE GOLD-BREASTED TRUMPETER.

But the Agami is quite a diſtinct race. It reſembles indeed the aquatic birds in the character which Adanſon has properly remarked, and alſo the greeniſh colour of the legs; but its nature is entirely different. It inhabits the arid mountains, and the upland foreſts; and never viſits the fens, or the margins of water.—We have here another example of the errors into which artificial ſyſtems lead.

Nor is it a Pheaſant or Curaſſo; for not only are its legs and thighs different, but its toes and nails are much ſhorter. Still more is it widely ſeparated from the Common Hen; and it cannot be ranged with the Cranes, ſince its bill, its neck, and its legs, are much ſhorter than in the aquatic birds.

The Agami is twenty-two inches long; its bill, which is exactly like that of the gallinaceous tribe, is twenty-one lines; its tail is very ſhort, not exceeding three inches and one-fourth, and is concealed by the ſuperior coverts, and does not project beyond the wings; its legs are five inches high, and completely covered with ſmall ſcales, as in the other gallinaceous birds, and theſe ſcales reach two inches above the knees, which are not feathered.

The whole of its head, its throat, and the upper half of its neck, both above and below, are covered with a ſhort down, which is very cloſe, and feels very ſoft; the fore-part of the lower ſurface of the neck, and the breaſt, are covered with a beautiful gorget four inches broad, whoſe brilliant colours vary between green, gold green, blue, and violet; the upper-part of its back, and the contiguous portion of its neck, are black; the plumage changes on the hind-part of the back into a tawny-rufous; but all the under-ſide of the body is black, and alſo the wings and the tail; only the great feathers which extend on the rump and the tail, are light aſh-coloured; the legs are greeniſh.

The nomenclators* have alſo confounded the Agami with the *Macucagua* of Marcgrave, which is the great *Tinamou*, and of which we ſhall treat in the following article, under the name of *Magua*. Adanſon is the firſt who detected this error.

Pallas † and Voſmaër ‡ have accurately aſcertained the ſingular power which this bird has of emitting

* Barrère, Briſſon, Voſmaër, &c.

† "The larynx, which without the breaſt is of the thickneſs of a ſwan's quill, and almoſt bony, grows much ſlenderer at its entrance into the breaſt, looſer, and cartilaginous, whence proceed two ſemi-cylindrical canals formed of membranes, and capable of extenſion.

"The air-bag on the right-ſide deſcends to the pelvis, and within the breaſt it is divided into three or four cells by tranſverſe membranous diaphragms. That on the left-ſide is much narrower, and terminates in the loins." *Miſcel. Zoolog.* p. 71

‡ The moſt characteriſtic and remarkable property of theſe birds conſiſts in the wonderful noiſe which they often make, either of themſelves, or when urged by the keepers of the *menagerie*. I do not wonder that hitherto they have been ſuppoſed to form this through the anus. It coſt me no little trouble to convince myſelf of the contrary. To ſucceed, one muſt be on the ground, and with a bit

emitting a dull hollow ſound, which was ſuppoſed to come from the anus*, and have diſco-

* M. de la Condamine entertained this opinion. *Voyage des Amazons*, p. 175.

a bit of bread entice the bird to come near; then make the noiſe, which the keepers can well imitate, and often diſpoſe the Agami to repeat it. This equivocal noiſe is ſometimes preceded by a ſavage cry, interrupted by a ſound approaching that of *ſcherck, ſcherck*, to which ſucceeds the hollow ſingular noiſe in queſtion, which reſembles ſomewhat the moan of pigeons. In this way it utters five, ſix, or ſeven times, with precipitation, a hollow noiſe emitted from within its body, nearly as if one pronounced *tou, tou, tou, tou, tou, tou*, with the mouth ſhut, reſting upon the laſt *tou* . . . a very long time, and terminating by ſinking gradually with the ſame note. This ſound alſo reſembles much the lengthened doleful noiſe which the Dutch bakers make, by blowing a glaſs trumpet, to inform their cuſtomers when the bread comes out of the oven. This ſound, as I have already ſaid, iſſues not from the anus; yet I am very confident, that it is formed by a ſlight opening of the bill, and by a ſort of lungs peculiar to almoſt all birds, though of a different form. This is alſo the opinion of M. Pallas, who heard it often with me, and to whom I gave one of the dead birds for diſſection." The doctor has communicated to me his obſervations with reſpect to the internal ſtructure of the animal, for which I am much obliged to him. "The wind-pipe," ſays he, "before its entrance into the breaſt, is as thick as a large writing-pen, bony, and quite cylindrical. In the breaſt it becomes cartilaginous, and divides into two ſemi-circular canals, which paſs through the lungs, the left one being very ſhort, but the right one reaching the bottom of the lower belly, and parted by tranſverſe membranes into three or four lobes."

Theſe lungs therefore are undoubtedly the inſtrument of the various cries emitted by birds. The air preſſed by the impulſive action of the fibres, ſeeks to eſcape through the large branches of the fleſhy lungs, and meets with an obſtruction from the little elaſtic membranes, which produces pulſations, the origin of all ſorts of ſounds †. But, what above all convinces us that this noiſe proceeds not from the anus, if a perſon obſerves attentively when the bird makes it, he will perceive the breaſt and belly to heave, and the bill to open ſomewhat. VOSMAER, *Amſterdam*, 1768.

† Memoires de l'Academie des Sciences, *année* 1753, p. 293.

vered

vered that this is a miſtaken notion. We ſhall only obſerve that in many birds, as well as in the Agami, the windpipe is bony at its opening, and becomes cartilaginous in its deſcent, and in general the cries of ſuch birds are deep; but there are alſo many birds on the other hand whoſe windpipe is cartilaginous at its riſe, and terminates bony in the breaſt, and theſe have commonly ſhrill notes.

The odd ſort of noiſe which this bird makes, is probably owing to the extent of lungs, and the capacity of their membranous cells. But it is unneceſſary to ſuppoſe with Voſmaër, that the Agami is obliged to open its bill a little in order to give paſſage to the ſound; for any ſudden motion in the bowels is communicated through the muſcles and teguments to the external air, which conveys the impulſe to the ear. We have often occaſion to notice this circumſtance; and it appears to be prejudice that the ſounds produced by animals are always tranſmitted through the throat, or through the alimentary canal. Nor is this ſpecies of ventriloquiſm peculiar to the Agami; the Curaſſo without opening its bill makes a ſimilar hollow ſound, which is even more articulate and more powerful. Indeed the ſame property ſeems to obtain, though in a leſs degree, in many kinds of birds in which the lungs are proportionally larger than in the quadrupeds. The hoarſe murmur which the Turkey-cock makes before his gobble, the cooing which the

the Pigeon effects without motion of the mouth, are of this nature; only in these the sound rises near the bottom of the throat; but in the Curasso, and especially in the Hocco, it has its origin deeper.

In regard to the manner in which the Agami lives in the domestic state, I shall quote the words of Vosmaër:—" When these birds are well kept, they are attentive to cleanliness, and often peck the feathers of the body and wings with their bill: if they frolic with each other, they perform all their movements by hopping, and violently flapping their wings. The change of food and of climate certainly cools here (in Holland) their natural ardor for propagation. Their ordinary subsistence is grain, such as buckwheat, &c. but they also eat readily small fish, flesh, and bread. This fondness for fish, and the uncommon length of their legs, shew that they partake of the nature of the Herons and Cranes, and that they belong to the class of the aquatic birds." We must observe here that the fondness for fish is no proof, since poultry are as greedy of this sort of food as of any other. " What *Pistorius* relates," continues Vosmaër, " with respect to the gratitude of this bird, may put many to the blush. When tamed, it distinguishes its master and benefactor with marks of its affection. Having reared one, I had an opportunity of experiencing this myself: when I opened its cage in the morning, the kind animal

mal hopped round me, expanding both his wings, and *trumpeting* (this is the term which we may employ to expreſs the noiſe) from his bill, and behind, as if he wiſhed me good morning. He ſhewed no leſs attention when I went out and returned again; no ſooner did he perceive me from a diſtance than he ran to meet me: and even when I happened to be in a boat, and ſet my foot on ſhore, he welcomed me with the ſame compliments, which he reſerved for me alone, and never beſtowed them upon others."

We ſhall ſubjoin a number of additional facts, which were communicated by M. de Manoncour.

In the ſtate of nature the Agami inhabits the vaſt foreſts in the warm climates of America, and never viſits the cleared grounds, ſtill leſs the ſettled ſpots. It aſſociates in numerous flocks, and prefers not the ſwamps and ſides of lakes; for it is often found on the mountains, and in hilly ſituations. It walks and runs rather than flies, ſince it never riſes more than a few feet, and only to reach ſome ſhort diſtance, or to gain ſome low branch. It feeds upon wild fruits, like the Curaſſos, the Marails, and other gallinaceous birds. When ſurpriſed in its haunts, it makes its eſcape by ſwiftneſs of feet, ſeldom uſing its wings, and at the ſame time emits a ſhrill cry like that of the Turkey.

Theſe birds ſcrape the earth at the roots of the large trees to form a bed for their eggs; and employ

employ no lining, and conſtruct no neſt. They lay many eggs, from ten to ſixteen; but the number is proportioned, as in all other birds, to the age of the female; they are almoſt ſpherical, larger than hens eggs, and tinged with light green. The young Agamis retain their down, or rather their firſt diſhevelled feathers, much longer than our chickens, or infant-partridges: theſe are ſometimes near two inches long, and before a certain age they might paſs for animals covered with ſilky hairs, which are cloſe like fur, and feel ſoft; the true feathers appear not till they have attained the fourth of their full growth.

The Agami is not only tamed eaſily, but becomes attached to its benefactor with all the fondneſs and fidelity of dogs; and of this diſpoſition it ſhews the moſt unequivocal proofs When bred up in the houſe, it loads its maſter with careſſes, and follows his motions; and if it conceives a diſlike to perſons on account of their forbidding figure, their offenſive ſmell, or of injuries received, it will purſue them ſometimes to a conſiderable diſtance, biting their legs, and teſtifying every mark of diſpleaſure. It obeys the voice of its maſter, and even anſwers to the call of all thoſe to whom it bears no grudge. It is fond of careſſes, and offers its head and neck to be ſtroked; and if once accuſtomed to theſe familiarities, it becomes troubleſome, and will not be ſatisfied without continual fondling. It makes

its

its appearance as often as its maſter ſits down to table, and begins with driving out the dogs and cats, and taking poſſeſſion of the room: for it is ſo obſtinate and bold, that it never yields, and often after a tough battle, can put a middle-ſized dog to flight. It avoids the bites of its antagoniſt by riſing in the air, and retaliates with violent blows with its bill and nails, aimed chiefly at the eyes; and after it gains the ſuperiority, it purſues the victory with the utmoſt rancour, and, if not parted, will deſtroy the fugitive. By its intercourſe with man, its inſtincts became moulded like thoſe of dogs; and we are aſſured the Agamis can be trained to tend a flock of ſheep. It even ſhews a degree of jealouſy of its rivals; for when at table it bites fiercely the naked legs of the negroes, and other domeſtics, who come near its maſter.

The fleſh of theſe birds, eſpecially when they are young, is not ill-flavoured, but is dry, and commonly hard. The rich brilliant part of the plumage which covers the breaſt, is ſeparated from the reſt, and prepared for the ornaments of dreſs.

M. De la Borde has alſo communicated the following particulars in regard to this bird. "The wild Agamis," ſays he, "are diſperſed "in the back country, and are no longer found "in the neighbourhood of Cayenne . . . and they "are very common in the remote unſettled "tracts. "They are always found in the im

"menſe

" menſe foreſts, in flocks from ten and twelve to " forty. . . . They fly from the ground to the " low trees, where they remain ſtill, and in " ſuch ſituation the hunters often kill ſeveral " without ſcaring away the reſt. . . . Some " perſons imitate their hoarſe murmur ſo exactly, " as to decoy them to their feet. . . . When the " hunters diſcover a flock of Agamis, they de- " ſiſt not till they have killed ſeveral: theſe birds " ſeldom or never fly, and their fleſh is but " ordinary, black, and always hard; however, " that of the young ones is more palatable. . . . " No bird is ſo eaſily tamed as this, and there " are always many of them in the ſtreets of " Cayenne. . . . They even roam out of town, " but return in due time to their maſter. . . . " They allow one to come near them, and handle " them at pleaſure; they are afraid neither of " dogs, nor of birds of prey, in the court-yard; " they aſſume the aſcendency over the poultry, " and keep them in great ſubjection: they feed " like the hens, the *Marils*, and the *Paraguas*; " but when very young, they prefer ſmall " worms and fleſh to every thing elſe.

" Almoſt all the birds have a trick of follow- " ing people through the ſtreets and out of town, " even perſons that they had never ſeen before. " It is difficult to get rid of them: if you enter " a houſe, they will wait your return, and again " join you, though often after an interval of " three hours." " I have ſometimes," adds M.

 de

de la Borde, " betaken myſelf to my heels, but " they ran faſter, and always got before me; " and when I ſtopped, they ſtopped alſo. I " know one which invariably follows all the " ſtrangers who enter its maſter's houſe; ac- " companies them into the garden, takes as " many turns. as they do, and attends them " back again *."

As the habits and œconomy of this bird were little known, I have thought proper to tranſcribe the different accounts which I have received. It appears that of all the feathered tribes, the Agami is the moſt attached to the ſociety of man; and in this reſpect it is as eminently diſtinguiſhed above them all, as the dog is above the other quadrupeds. The diſpoſition of the Agami is the more remarkable, ſince it is the only bird that has a ſocial turn; whereas ſeveral of the quadrupeds diſcover attachment to man, though inferior in degree to that of the dog. And is it not ſtrange, that an animal, ſo peculiarly formed for ſociety, has never been domeſticated? Nothing can better ſhew the immenſe diſtance between the civilized man and the rude ſavage, than the dominion obtained over the lower creation. The former has made the dog, the horſe, the ox, the camel, the elephant, the rein-deer, &c. ſubſervient to his utility, or his pleaſure: he has drawn together

* Note communicated by M. de la Borde, King's phyſician at Cayenne, in 1776.

the

the hens, the geese, the turkies, and the ducks, and has lodged the pigeons. The savage has overlooked advantages the most obvious and the most essential to his comfort. It is society that gives spring to activity; that awakens the dormant faculties; and that expands, informs, and enlivens the whole! [A]

[A] Specific character of the Gold-breasted Trumpeter:—"Its head and breast are smooth and shining green."

The TINAMOUS*.

THESE birds, which are peculiar to the warm parts of America, may be regarded as a part of the gallinaceous claſs; for they reſemble the Buſtard and Partridge, though they differ in ſeveral properties. But there are certain habits in animals which reſult from the nature of the climate, and from local circumſtances, and which ought not to be deemed eſſential characters.—Thus many birds, ſuch as Partridges, which remain conſtantly on the ground in Europe, perch in America; and even the palmated aquatic fowls, paſs the day in the water, and return to lodge during the night among the trees. The dangers with which they are ſurrounded appear to drive them to ſuch retreats. The immenſe ſwarms of inſects and reptiles, engendered by the heat and the moiſture of the climate, threaten every moment their deſtruction. If they ventured to repoſe upon the ground, the denſe columns of ants would attack them in their ſlumbers, and reduce them to ſkeletons. The Quails are the only birds in thoſe countries which reſt upon the ſurface; and they often fall a prey to the voracity of the ſerpents. Nor is it im-

* This is the name given to theſe birds in Guiana.

probable,

probable, that the Quails have been introduced ſince the diſcovery of America, and that they have not yet acquired the habits ſuited to their new ſituation, or learned to guard againſt the aſſaults of their numerous foes.

We ſhould have ranged the genus of the *Tinamous* after that of the Buſtard ; but theſe birds were, at that time, but imperfectly known, and we are indebted to M. de Manoncour for the principal facts relating to their hiſtory, and alſo for the ſpecimens preſented to the Royal Cabinet, from which we have made the deſcriptions.

The Spaniſh inhabitants of America *, and the French ſettlers at Cayenne, have both termed theſe birds *Partridges ;* and the appellation has been adopted by ſome nomenclators †, though altogether improper : for the *Tinamous* are diſtinguiſhed by their long ſlender bill, blunt at the tip,. black above, and whitiſh below ; their noſtrils oblong, and placed near the middle of their bill ; their hind-toe is very ſhort, and does not reſt upon the ground ; their nails are very ſhort, broad, and channelled beneath ; their legs alſo differ from thoſe of Partridges, being covered behind, as in the poultry, with ſcales, their whole length, ſhaped like ſmall ſhells ; but the upper-part projects and forms inequalities not obſerved on the legs of poultry. In all the Tinamous,

* Letter of M. Godin des Odonnais, to M. de la Condamine, 1773, p. 19. note firſt.

† Briſſon.—Barrere.

 the

the throat and craw are thinly ſtrewed with ſtraggling feathers; the quills of the tail are ſo ſhort, that in ſome they are wholly concealed by the ſuperior coverts.—Thus they are improperly named Partridges, ſince they differ in ſo many eſſential characters.

But they differ alſo from the Buſtard, by ſeveral of their principal characters, and eſpecially by having a fourth toe behind, which is wanting in the Buſtard. In ſhort, we have judged it requiſite to range them in a ſeparate genus, under the name which they receive in their native country.

All the ſpecies of the Tinamous paſs the night upon the trees, and ſometimes perch during the day; but they always ſettle among the loweſt branches, and never mount to the ſummits: and this circumſtance ſeems to imply the probability that they are not actuated by original impulſe, but directed by conſiderations of ſafety.

The Tinamous are, in general, excellent for the table; their fleſh is white, firm, cloſe, and juicy, eſpecially about the wings, and taſtes like that of the Red Partridge. The thighs and rump have commonly a diſagreeable bitterneſs, which is occaſioned by the fruit of the Indian reed upon which they feed. The ſame bitter taſte is obſerved in the Ring-Pigeons which eat theſe fruits. But when the Tinamous live upon other fruits, ſuch as wild cherries, &c. their fleſh is uniformly delicate, but ſtill has none

of

of the *fumet*. In the ſultry humid climate of Cayenne, meat will not keep more than twenty-four hours from putrefaction, and no ſort of game can be allowed time to mellow and acquire that delicious flavour which conſtitutes its excellence. Theſe birds, like all thoſe which have a craw, often ſwallow the fruits without bruiſing or even cracking them; they are particularly fond of the wild cherries, and alſo of the produce of the *common palm*, and even of that of the coffee-ſhrub, when they can find it. Nor do they cull their ſubſiſtence from the trees; they only collect the fruits which have dropped. They ſcrape the ground to form their neſt, which is uſually nothing but a ſingle layer of dry herbs. They lay twice a-year, and have numerous broods; which ſhews that theſe birds and the Agamis are of the gallinaceous claſs, which is remarkably prolific. Like theſe alſo, they fly heavily, and to ſhort diſtances, but run ſwiftly on the ground; they form little flocks, and it is uncommon to find them either ſingle or in pairs; they call each other in all ſeaſons; in the morning and evening, and ſometimes too during the day: this call is a ſlow, quavering, plaintive whiſtle, which the fowlers imitate to bring them near; for this game is the moſt common and the beſt which that country affords.

We ſhall add a remarkable circumſtance with reſpect to theſe genus of birds, that, as in the Anters, the female is larger than the male; a

 property

property which in Europe is found only in the rapacious tribe. In the ſhape of the body, however, and in the diſtribution of the colours, the females are almoſt entirely like the males.

The GREAT TINAMOU.

Le Magoua, Buff.
Tinamus Braſilienſis, Lath. Ind,
Tetras Major, Gmel.
Perdix Braſilienſis, Briſſ.
Macucagua, Ray, Will. and Klein.

Firſt Species.

This bird is as large as a Pheaſant, and, according to Marcgrave, it has twice as much fleſh as a plump hen *. The throat and the lower-part of the belly are white; the upper-part of the head is deep rufous; the reſt of the body is of a brown-gray variegated with white on the top of the belly, the ſides, and the coverts of the thighs; there is a little greeniſh on the neck, the breaſt, the riſe of the back, and the ſuperior coverts of the wings and of the tail, on which ſome blackiſh tranſverſe ſpots are obſerved, that are leſs numerous on the coverts of the tail; the brown-gray is deeper on the reſt of the body, and variegated with black tranſverſe ſpots,

* This bird eats, according to that author, wild beans, and the fruit of a tree called in Brazil, *araeicu*.

which

N.º 111

THE GREAT TINAMOU.

which are leſs frequent near the rump: there are alſo ſome ſmall black ſpots on the lateral quills of the tail; the middle quills of the wings are variegated with rufous and brown-gray, and terminated by a ruſty border; the great quills are cinereous, without any ſpots or border; the legs are blackiſh, and the eyes black, and a little behind them the ears are placed, as in the poultry. Piſo remarks, that the internal ſtructure of this bird is exactly like that of the hen.

The ſize varies in different ſubjects; the average meaſures are, total length fifteen inches, the bill twenty lines, the tail three inches and a half, the legs two inches and three-fourths; the tail projects an inch and two lines beyond the wings.

The call of the Great Tinamou is a hollow ſound, which may be heard at a great diſtance, and is whiſtled preciſely at ſix o'clock in the evening, the time when the ſun ſets in that latitude. It is ſilent during the night, unleſs it be ſcared.

The female lays twelve or fifteen eggs, which are almoſt round, rather larger than hens eggs, of a beautiful greeniſh blue, and are excellent eating. [A]

[A] Specific character of the *Tinamus Braſilienſis*, Lath.:— "It is duſky-olive, ſpotted with duſky, its belly whitiſh and variegated, the thighs rough behind."

The CINEREOUS TINAMOU.

Tinamus Cinereus, Lath. Ind.
Tetrao Cinereus, Gmel.

Second Species.

The epithet *cinereous* will ſerve for a deſcription of this bird; for that colour is uniform over the whole body, except a tint of rufous on the head and the top of the neck. It has the ſame ſhape as the other, only it is ſmaller. It is a new ſpecies communicated by M. de Manoncour. It is of all the Tinamous the leaſt frequent in Cayenne.

Its length is a foot; its bill ſixteen lines; its tail two inches and a half; and its legs the ſame. [A]

[A] Specific character of the *Tinamus Cinereus*: LATH.:—" It is cinereous-brown, its head and neck tawny."

The VARIEGATED TINAMOU.

Tinamus Variegatus, Lath. Ind.
Tetrao Variegatus, Gmel.

Third Species.

This ſpecies, which is the third in the order of ſize, differs from the two firſt by its variegated plumage.

plumage. The Creoles of Cayenne call it the *Pintado Tinamou*; but this appellation is improper, for it bears no reſemblance to the Pintado, and its ſtriped plumage is not dotted. Its throat and the middle of its belly are white; its tail, its breaſt, and the top of its belly, rufous; its ſides and its thighs ſtriped obliquely with white, with brown, and with rufous; the upper-part of its head, and the top of its neck, black; all the upper-part of its body, the ſuperior coverts of its tail and of its wings, and the middle quills of its wings ſtriped tranſverſely with black and olive brown, deeper on the back, and lighter on its rump and on its flanks; the great quills of its wings are brown, and uniform without a ſpot; its legs are blackiſh.

Its total length is eleven inches; its bill fifteen lines; its tail two inches, and exceeds the wings by ſix lines.

It is pretty common in Guiana, though not ſo numerous as the Great Tinamous, which occur indeed the moſt frequently in the woods, for none of theſe three ſpecies haunt the cleared ground. The female Variegated Tinamou lays ten or twelve eggs, which are rather ſmaller than thoſe of the hen Pheaſant, and are uniformly tinged with a beautiful black. [A]

[A] Specific character of the *Tinamus Variegatus* of LATH.:—
" It is ſtriped with rufous, brown, and blackiſh; below rufous,
" with a black cap; its throat, and the middle of its belly,
" white."

The

The LITTLE TINAMOU.

Le Soui, Buff.
Tinamus-Soui, Lath. Ind.
Tetrao-Soui, Gmel.

Fourth Species.

Soui is the name by which this bird is known in Guiana, and which was given by the natives of the country. It is the ſmalleſt of the genus, not exceeding eight or nine inches in length, and not being larger than a Partridge. Its fleſh is as delicate as that of the other kinds, but it lays only five or ſix eggs, and ſometimes no more than three or four, which are rather larger than Pigeons eggs; they are almoſt ſpherical, and as white as thoſe of hens. The Little Tinamous do not form their neſt like the Great Tinamous, by ſcraping the ground; they build it with long narrow leaves on the loweſt branches of ſhrubs: it is hemiſpherical, about ſix inches in diameter, and five inches high. Of all the four ſpecies, this is the only one which does not live conſtantly in the woods; it often frequents the young ſtraggling trees and buſhes, which ſhoot up in land that has been cultivated and abandoned; and ſometimes it even viſits dwellings.

Its

Its throat is variegated with white and rufous; all the under-part of its body and the coverts of its thighs are of a light-rufous; the upper-part of its head and of its neck are black; the lower-part of its neck, its back, and all the under-part of its body, are brown, radiated with dull blackiſh; its ſuperior coverts and the middle quills of its wings are brown, edged with rufous; the great quills of its wings are brown, without any ſpots or borders; its tail projects ten lines beyond its wings, but is exceeded by its own coverts. [A]

[A] Specific character of the *Tinamus-Soui* of LATH.—" It is clouded with brown; below rufous; its throat variegated " with white; the upper-part of its head and the hind-part of its " neck, black."

The TOCRO.

Tetrao Guianensis, Gmel.
Perdix Guianensis, Lath. Ind.
The Partridge of Guiana, Buff.

THE Tocro is somewhat larger than our Gray Partridge, and its plumage deeper: but in other respects, it is exactly similar; in its figure, in the proportions of its body, in the shortness of its tail, and in the shape of its bill and legs. The natives of Guiana call it *Tocro*, a name which well expresses its cry.

These Partridges of the New World have nearly the same habits as those of Europe; only they still continue in the forests, because they have not been accustomed to cleared grounds. They perch on the low branches of the bushes, but only to pass the night; which is to avoid the damp, and perhaps the swarms of insects. They lay commonly twelve or fifteen eggs, which are entirely white; the flesh of the young ones is excellent, but has no *fumet*. The old ones are also eaten, and are even more delicate than ours; but the rapid progress of putrefaction in those climates will not allow sufficient time for acquiring the proper flavour.

As

As our Gray Partridges do not intermix with our Red Partridges, it is highly probable that the Brown Partridges of America would breed with neither, and consequently are a separate species. [A]

[A] Specific character of the *Perdix Guianensis* of LATH.— "It is rufous-brown, spotted and variegated; its throat is cinereous; a fulvous fillet passes over its eyes; its wing-quills are marked externally with rufous spots."

The FLYCATCHERS.

Les Gobe-Mouches, Moucherolles & Tyrans, Buff.

NATURE has aſſigned theſe a place after the humbleſt of the rapacious tribes. They are harmleſs and even uſeful; they conſume not fruits, but live upon flies, gnats, and other winged inſects. The genus comprehends numerous ſpecies, which vary exceedingly in point of ſize, from that of the Nightingale to that of the Shrike. Some characters however are common to them all: their bill is compreſſed, broad at the baſe, and almoſt triangular, beſet with briſtles, and the tip bent into a little hook in many of the middle ſpecies, and more curved in all the large ſpecies; the tail is of conſiderable length, and only half covered by the wings. Their bill is alſo ſcalloped near the point; a property which they ſhare with the Blackbird, the Thruſh, and ſome other birds.

Their diſpoſition is in general ſhy and ſolitary, and their notes are neither lively nor melodious. Subſiſting in the region of air, they ſeldom leave the ſummits of the lofty mountains, and are rarely ſeen on the ground. Their habit of clinging to the branches would ſeem to have increaſed

the

the growth of their hind-toe, which in moſt of the Flycatchers is longer than the fore-toe.

The ſultry tropical countries, which teem with various inſects, are the favourite abodes of theſe birds *. Two ſpecies only are found in Europe; but we reckon eight in Africa, and in the warm regions of Aſia, and thirty in America, which are alſo the largeſt ſpecies; and as in the New World the inſect nations are the moſt numerous and the moſt formidable, ſo Nature has provided a ſtronger body to prey upon them. —We ſhall range them according to their ſize into three diviſions: the firſt are ſmaller than that of the Nightingale, and are the *Flycatchers* properly ſo called; the ſecond are ſomewhat larger, and may be termed *Moucherolles*; the third are the *Tyrants*, which are nearly as large, if not larger than the Wood-chat, or Rufous Shrike, and reſemble in their ſhape and inſtinct the genus of the Shrikes, which ſeems to connect the claſs of rapacious birds with the Flycatchers.

* "The Flycatchers are in general common birds in hot countries. The ſpecies are there more frequent and more numerous than in temperate countries; and few occur in cold climates. They feed only upon inſects. They are deſtructive creatures, which, in the hot and moiſt regions, Nature has oppoſed to the exceſſive fecundity of the inſects." *Voyage à la Nouvelle Guinée*, par M. SONNERAT.

The SPOTTED FLYCATCHER.

Le Gobe-Mouche, Buff.
Muſcicapa-Griſola, Linn. and Gmel.
Griſola, Aldrov.
Sylvia Fuſca, Klein.
Muſcicapa, Briſſ.
The Cob-web, Mort. Northamp.

Firſt Species.

This ſpecies being well known, will ſerve as a term of compariſon.—It is five inches eight lines in length; its alar extent eight inches and a half; the wings, when cloſed, reach to the middle of its tail, which is two inches long; its bill is flat, broad at its baſe, and eight lines long, beſet with briſtles: its whole plumage conſiſts of theſe three colours, gray, white, and blackiſh cinereous; its throat is white; its breaſt and the ſides of its neck are ſpotted with faint ill-defined brown; the reſt of the under-part of its body is whitiſh; the upper-part of its head appears variegated with gray and brown; all the higher-part of its body, its tail, and its wings, are brown; the quills and their coverts are lightly fringed with whitiſh.

The ſpotted Flycatchers arrive in April, and depart in September. They live generally in the foreſts, and prefer the ſolitude of the cloſe ſhady ſpots; and ſometimes they are found in the

N.° 112

FIG. 1. THE SPOTTED FLY-CATCHER OF LORRAIN.
FIG. 2. THE COMMON SPOTTED FLY-CATCHER.

thick vineyards. They have a melancholy air; their diſpoſition is wild, inanimate, and even ſtupid: they place their neſt, entirely expoſed, either on the trees or the buſhes. No ſmall bird is ſo incautious, and none has inſtincts ſo unſettled. The neſts are not of an uniform conſtruction; ſome conſiſt entirely of moſs, and others have a mixture of wool. They conſume much time and labour upon the rude ſtructure, and ſometimes we find it interwoven with thick roots, and are ſurprized that ſo ſmall an artificer could employ ſuch materials. They lay three or four eggs, ſometimes five, which are covered with rufous ſpots.

Theſe birds procure the principal part of their ſubſiſtence while on the wing, but ſeldom alight, and then only by ſtarts, upon the ground, and never run along it. The male is not different from the female, except that its face is more variegated with brown, and its belly is not ſo white. They appear in France in the ſpring, but the cold weather which ſometimes prevails in the middle of that ſeaſon is pernicious to them. Lottinger obſerves, that they almoſt all periſhed in the ſnows which fell in Lorraine in April 1767 and 1772, and that they were caught by the hand. Every degree of cold that deſtroys the inſects, their only ſupport, muſt prove fatal to them; accordingly they leave our provinces before the froſt ſets in, and they are never ſeen after the end of September. Aldrovandus ſays, that they do not migrate; but this muſt be underſtood in regard to Italy, or of ſtill warmer countries. [A]

[A] Specific character of the Spotted Flycatcher, *Muſcicapa-Griſola*:—" It is duſkiſh, below whitiſh, its neck ſpotted longi-" tudinally, its vent tawny." The Flycatcher appears in England in the ſpring, and retires in Auguſt. " It is of all our ſummer " birds," ſays Mr. White, " the moſt mute, and the moſt familiar " It builds in a vine, or a ſweetbriar, againſt the wall of an houſe, " or in the hole of a wall, or on the end of a beam or plate, " and often cloſe to the poſt of a door where people are going in " and out all day long. This bird does not make the leaſt " pretenſion to ſong, but uſes a little inward wailing note, when " it thinks its young in danger from cats or other annoyances: " it breeds but once, and retires early." When its young are able to fly, it retires with them to the thick woods, and frolics among the high branches, ſinking and riſing often perpendicularly in queſt of flies, which hum below.

The COLLARED BLACK FLYCATCHER, *or, the* FLYCATCHER *of* LORRAINE.

Muſcicapa Atricapilla, var. Linn.
The Red Flycatcher, var. Lath.

Second Species.

It appears to be better known in Lorraine, and more common than in other parts. It is rather ſmaller than the preceding, being ſcarcely five inches long; it has no other colours than white and black, which are diſperſed in diſtinct ſpots; but its plumage notwithſtanding varies more remarkably than that of any other bird.

The

The male appears to have four different garbs, according to the ſeaſons. The firſt is that of the autumn, or winter, when the plumage is the ſame with that of the female, which is not ſubject to ſuch changes. The ſecond is worn when theſe birds arrive in Provence or Italy, and is then exactly like that of the Epicurean Warbler. The third is what they aſſume ſhortly after their appearance, and may be termed the ſpring attire *. This is only the intermediate gradation to the fourth ſtate, which is that of ſummer, and which, as Lottinger obſerves, may be properly termed its *marriage ſuit*, becauſe it aſſumes this at pairing, and lays it aſide after the breeding is over. The bird is then in its full beauty: a white collar three lines broad encircles its neck, which is of the fineſt black; its head is of the ſame colour, except the front and the face, which are bright white; its back and its tail are ſtained with the black of the head; its rump is variegated with black and white; a white ſtreak of a line in breadth borders for ſome way the outermoſt quills of the tail; the wings, conſiſting of ſeventeen quills, are of a deep cheſnut; the third and the four following are tipt with a much lighter brown, which, when the wings are cloſed, has a very fine effect: all the quills, the two firſt excepted,

* "I fed one this ſpring three or four days. Every perſon admired it, though one of its fineſt ornaments (the collar) was wanting. The white and black of its plumage were of the brighteſt tinte" *Letter of* M. LOTTINGER, 30th *April* 1772.

 have

have a white ſpot on the outer edge, which enlarges the nearer it is to the body; ſo that the outer-edge of the laſt quill is entirely white; the throat, the breaſt, and the belly are white; the bill and the legs black. There is a remarkable luſtre and gloſs ſpread over the whole of the plumage; but theſe beauties are gone before the beginning of July. The colours grow dilute and duſky; the collar firſt diſappears, and the reſt ſoon becomes ſtained and obſcure, and the male is no longer diſtinguiſhable from the female. "I have frequently met with bird-catchers," ſays Lottinger, "who ſpread the nets on the ſprings "in places where they breed; and though it "was only in July, they told me that they caught "numbers of females, but not a ſingle male;" ſo entirely was the external diſtinction of ſex obliterated. That naturaliſt has not deſcribed ſo fully the vernal plumage with which they enter into the ſouthern provinces. However, Aldrovandus ſeems to indicate the change of this Flycatcher, which he has well deſcribed in another place *, when, ranging it again with the Becafigos, he tells us of his having ſurprized it at the very inſtant of its transformation, being then neither an *Epicurean Warbler*, nor a *Black-Cap*. Already, he ſubjoins, the collar was become white; there was a white ſpot on the front;

* He deſcribes its collar, the white ſpot on its wing: he commends its beauty. It is known, he ſays, by the fowlers of Bologna, under the name of *Peglia-Moſche*.

white

white on the tail and the wings; the under-part of the body white, and the reſt black. Theſe properties ſufficiently diſcriminate the Collared Black Flycatcher.

This bird arrives in Lorraine about the middle of April. It lives in the foreſts, thoſe eſpecially which conſiſt of tall trees, and breeds in the holes of the trunks, ſometimes pretty deep, and at a conſiderable height above the ſurface of the ground. Its neſt is formed of ſmall ſtalks of graſs, and a little moſs, which covers the bottom of the cavity. It lays ſix eggs. After the young are hatched, the parents frequently go in and out, carrying ſupplies of food; and this care of their infant brood often betrays the retreat, which would otherwiſe be difficult to diſcover.

They ſubſiſt only upon flies, and other winged inſects. They are never ſeen on the ground; and for the moſt part they keep very high, fluttering from tree to tree. They have no ſong, but only an exceeding ſhrill plaintive accent, which turns upon the ſharp note *crrî, crrî*. They appear ſad and gloomy; but their attachment to their offspring inſpires activity, and even courage.

Lorraine is not the only province in France where the Collared Black Flycatcher is found. Hebert has informed us, that one was ſeen in Brie, but where it is little known, becauſe it is wild and tranſitory. We ourſelves found one of theſe Flycatchers on the tenth of May 1773, in a ſmall park near Montbard in Burgundy; and it was in the ſame ſtate of plumage as that deſcribed by Briſſon. Of the great coverts, which he ſays are tipt with white, thoſe only which were next the body were ſuch, and the more remote were brown; the inferior coverts alone of the tail were white, the ſuperior ones were blackiſh-brown; the rump was dull pearl-gray; the nape of the neck where the collar was ſituated, was lighter than the head and the back; the middle quills of the wings were near the tip of the ſame brown as the great quills; the tongue appears to be indented at the tip, broad for the ſize of the bird, but proportioned to the breadth of the bottom of the bill; the inteſtinal tube was eight or nine inches long; the gizzard muſcular, preceded by a dilatation of the *œſophagus;* there were ſome marks of a *cæcum;* and no gall-bladder. The bird was a male, and the teſticles ſeemed to be a line in diameter; it weighed three gros.

In this ſpecies of Flycatchers, the ends of the wings meet, and ſtretch beyond the middle of the tail; which is the reverſe of what generally takes place in the genus.— There are ſeveral inaccuracies in the figures given of it in the *Planches, Enluminées.*

This penſive bird enjoys a quiet peaceful life, protected by ſolitude. It avoids the cold ſeaſon, and ſhifts the ſcene to the genial climes of the ſouth,

ſouth, there to renew its loves. They are found, however, pretty far north, ſince they inhabit Sweden*. — There are two ſpecies from the Cape of Good Hope, which ſeem to be the ſame with that of Lorraine; the firſt, being diſtinguiſhed only by a ruſty ſpot on the breaſt; and the ſecond is only the female. The difference of appearance is very ſlight, if we eſtimate the influence of ſo diſtant a climate.

* Fauna Suecica.

The UNDULATED FLYCATCHER.

Le Gobe-Mouche de l'Ile de France, Buff.
Muſcicapa Undulata, Gmel.

Third Species.

We have in our cabinet two Flycatchers ſent from the Iſle of France; the one rather black than brown, and the other ſimply brown. Both are ſmaller, and eſpecially ſhorter, than the European Flycatchers. In the firſt, the head is blackiſh-brown, and the wings ruſty-brown; the reſt of the plumage is a mixture of whitiſh and of a brown, like that of the head and wings, diſpoſed in ſmall waves, or ſmall ſpots, without much regularity. — The ſecond appears to be only the female of the firſt. In fact, the differences are too ſlight to conſtitute two ſpecies; for the ſize, the figure, the colours, and almoſt the ſhades are ſimilar. The ſecond has indeed more white, mixed with ruſty on the breaſt and belly; the brown-gray on the head and body is more dilute; but the colours of the female are lighter in all the ſpecies of birds. [A]

[A] Specific character of the *Muſcicapa Undulata*:— "It is "waved with whitiſh and brown; its head partly blackiſh; its "wings duſky rufous."

The SENEGAL FLYCATCHER.

Le Gobe-Mouche à Bandeau Blanc du Senegal, Buff.
Muſcicapa Senegalenſis, Linn. and Gmel.

Fourth Species.

Under this appellation we ſhall comprehend the two birds figured in the *Planches Enluminées*, by the names of *Rufous-breaſted Flycatcher of Senegal*, and *Black-breaſted Flycatcher of Senegal*. Theſe handſome birds may be deſcribed together; they are of the ſame ſize, and are natives of the ſame climate; and the diſtribution of their colour is ſimilar in both. It is probable that they are the male and female of the ſame ſpecies. The white line which paſſes upon the eye, and encircles the head with a ſort of little diadem, is not ſo entire or diſtinct in any other of the genus. The firſt is the ſmaller, being only three inches and a half long; a rufous ſpot covers the top of the head, which is ſurrounded by the white

white ring : from the exterior angle of the eye an oval black ſpot extends, which is bounded above by the ring, and ſtretches into a point near the tip of the bill ; the throat is white ; a light rufous ſpot marks the breaſt ; the back is light-gray, ſpread upon white ; the tail and the wings are blackiſh. A white line extends obliquely on their middle coverts, and the ſame coverts are edged with ſcales of the rufous colour of the breaſt. A gloſſy tranſparency is ſpread over all the plumage of this bird ; ſtill lighter and more vivid on that of the other, which is ſimpler in its colours, conſiſting of a mixture of light gray, of white, and of black, and is not inferior in point of beauty ; the white bar paſſes upon the eyes ; a horſe-ſhoe of the ſame colour riſes pointed under the bill, and is cut ſquare on the breaſt, which is diſtinguiſhed by a black belt ; the top of the neck is black, which mingling with the white of the back melts into gray ; the quills are black, fringed with white, and the white line of the coverts opens into feſtoons ; the ſhoulders are black ; but there is a little fringed white interwoven with all this black ; and through all the white of the plumage ſmall black ſhades are interſperſed, which are ſo light and tranſparent, that this little bird is more beautiful than many which are decorated with a profuſion of rich and vivid tints. [A]

[A] Specific character of the *Muſcicapa Senegalenſis* :—" It is " variegated ; its eye-brows are white ; the outermoſt tail-quills " are white one half of their length."

The

The BOURBON FLYCATCHER.

Le Gobe-Mouche Huppé du Senegal, Buff.
Muſcicapa Borbonica, Gmel.

Fifth Species.

We ſhall conſider the Creſted Flycatcher of the iſland of Bourbon as only a variety of the Creſted Flycatcher of Senegal, and both as forming one ſpecies. The iſland of Bourbon, placed in the midſt of a vaſt ocean, and ſituated between the tropics, enjoys an uniform temperature, which requires not periodical migrations, and when firſt viſited by the European ſhips contained no land bird. Thoſe found in it at preſent have been carried thither by chance or deſign; nor muſt it be regarded as the native ſeat of original ſpecies*: we ſhall therefore claſs

* We find alſo two Flycatchers of the iſle of Bourbon, which we ſhall barely mention, convinced that they belong to ſome ſpecies on the continent of Africa. The one is repreſented in the *Illumined Plates*, N° 572, Fig. 3; it is ſmall, and quite black, except a little rufous which it has under its tail; and, notwithſtanding the difference of colour, we may ſuppoſe it to be a variety of the Cape Flycatchers, which we have already referred to our Collared Black Flycatcher: theſe differences of plumage being apparently no other than what we ſee it undergo itſelf, and which the influence of a hotter climate muſt render more extenſive and rapid, eſpecially as it is naturally diſpoſed to change. M. Briſſon indicates in the following terms the third Flycatcher of the iſle of Bourbon, to which he ſays the inhabitants give the name of *Tecteo*:—"Flycatcher, "above brown; the edges of the quills tawny; below tawny; "*(male)*. Dirty white; the tail-quills deep brown; their outer "edges light brown; *(female)*."

the

the bird discovered on the island with its analogous one of the continent. In fact, the differences between them are not greater than those which often occur among individuals of this genus; their figure, their bulk, and their principal colours, are the same; in both the head is furnished with small feathers, half-raised into a black crest, with green and violet reflections; this black descends in the Senegal Flycatcher like a square spot upon the breast, and the forepart of the neck. In that of Bourbon, the black covers only the head, with the eye, and also the lower-mandible; but, in other subjects, it is spread also upon the top of the neck. In both the under-part of the body is of a fine light slate-gray, and the upper-side bay, which is more vivid in that of Bourbon, and deeper and chesnut in that of Senegal; and this colour, which extends equally over the whole of the tail and wings of the last, is intersected by a little white in the other, and assumes a deeper cast on the coverts, which are also fringed with three lighter streaks. The blackish colour of the quills has only a light rusty border on the outside, and whitish on the inside of the webs. The greatest difference occurs in the tail; that of the Bourbon Flycatcher is short and square, being only two inches and a half long; the tail of the Senegal Flycatcher is more than four inches, and is tapered from the two middle quills, which are the longest, to the outer ones, which are two inches

inches ſhorter. This difference may be imputed to the effect of age, ſeaſon, or of ſex: at any rate, the claſſing of them together will excite a fuller inveſtigation, and an attention to the points of diſcrimination.

The BROWN-THROATED SENEGAL FLYCATCHER.

Muſcicapa Melanoptera, Gmel.
The Collared Flycatcher, Lath.

Sixth Species.

This Flycatcher was brought from Senegal by Adanſon. It is the ſame with what Briſſon deſcribes under the appellation of *Collared Senegal Flycatcher*, which is improper, ſince neither the brown ſpot on the throat, nor the black line that bounds it, can be termed a collar. A brown cheſnut ſpot riſes with a ſtraight tranſverſe margin under the bill and the eyes, and ſpreads on the throat, but extends not to the breaſt, being terminated at the lower part of the neck with a narrow black line, which is very diſtinct, as the breaſt, with the reſt of the lower-part of the body, is white; the upper ſurface is of a fine bluiſh gray; the tail blackiſh; the outermoſt quill is white on the outſide; the great coverts of the wings are white alſo, the ſmall ones blackiſh; the quills are deep cinereous, fringed

fringed with white, and the two next the body are white through their outer half; the bill is broad and flat, and beset with bristles at the angles. [A]

[A] Specific character of the Collared Flycatcher, *Muscicapa Melanoptera*, GMEL.—" It is cinereous; below white; the throat " tawny-bay; a transverse black belt below; the bill, wings, and " tail, black."

The AZURE FLYCATCHER.

Le Petit Azur, Gobe-Mouche Bleu des Philippines, Buff.
Muscicapa Cærulea, Gmel.

Seventh Species.

A beautiful azure covers the back, the head, and all the fore-part of the body of this pretty Flycatcher, except a black spot on the back of the head, and another black spot on the breast; the blue extends to the tail, and gradually grows more dilute; it tinges the small webs of the wing-quills, of which the rest are blackish; and it also gives shades to the white of the ventral feathers.

This bird is rather smaller, taller, and slenderer, than the Spotted Flycatcher of Europe. Total length five inches; the bill seven or eight lines, and not scalloped or hooked; the tail two inches, slightly tapered; the blue has a glossy lustre.

The

The WHISKERED FLYCATCHER.

Le Barbichon de Cayenne, Buff.
Musticapa Barbata, Gmel.

Eighth Species.

In all the Flycatchers the bill is beſet with briſtles; but, in the preſent, theſe are ſo long that they reach to the tip, which is the reaſon of its epithet *whiſkered.* It is near five inches long; its bill very broad at the baſe, and very flat through its whole length; the upper mandible projects a little beyond the lower, all the upper-ſide of the body is deep olive-brown, except the top of the head, which is covered by orange feathers, partly concealed by the other feathers; the under-ſide of the body is greeniſh-yellow, which, on the rump, runs into a fine yellow.

The female is ſomewhat larger than the male; all the upper-ſide of its body is blackiſh-brown, mixed with a ſlight tint of greeniſh, not ſo conſpicuous as in the male; the yellow of the top of the head forms only an oblong ſpot, which is partly hid by the feathers of the general colour; the throat and the top of the neck are whitiſh; the feathers of the reſt of the neck, of the breaſt, and of the under-ſurface of the wings, have their middle brown and the reſt yellowiſh;

the

the belly and the under-surface of the tail are entirely of a pale yellow; the bill is not so broad as in the male, and has only a few short bristles on each side.

The notes of the Whiskered Flycatcher are not sharp; it whistles gently the sound *pipi;* the male and female generally keep together. The incautious manner in which the Flycatchers place their nest is remarkable in this species; it does not seek the leafy boughs, but builds on the most naked and exposed branches. The nest is the more easily detected, as it is exceedingly large, being twelve inches high, and more than five in diameter, and entirely composed of moss; it is closed above, and has a narrow aperture in the side, three inches from the top. We owe our information to M. de Manoncour. [A]

[A] Specific character of the Whiskered Flycatcher, *Muscicapa Barbata*, GMEL:—"It is olive-brown, below greenish-yellow, "its top orange, its rump yellow."

The BROWN FLYCATCHER.

Le Gobe-Mouche Brun de Cayenne, Buff.
Muscicapa Fuliginosa, Gmel.

Ninth Species.

The Brown Flycatcher is scarcely four inches long; the feathers of its head and back are blackish-brown, edged with fulvous brown; the fulvous is deeper, and predominates on the quills of the wing, and the black on thoſe of the tail, which have a whitiſh fringe; all the under-ſide of the body is whitiſh, except a fulvous tint on the breaſt; the tail is ſquare, and half covered by the wings; the bill is ſharp, with ſmall briſtles at its root;—ſuch are the diſtinguiſhing features of this little bird. Its ſpecies ſeems however to admit a variety, if the differences which we perceived in another ſubject are not to be imputed to age or ſex. The duſky ground of the plumage, in this laſt bird, had a yellowiſh tint under the belly, and an olive-brown on the breaſt; the head and back had a ſlight caſt of a deep olive-green, and on the great quills of the wings were ſome lighter ſtreaks, but the ſmall coverts were dyed with a pale light roſe-yellow. [A]

[A] Specific character of the Brown Flycatcher, *Muſcicapa Fuliginoſa*, GMEL.: — "It is dark brown, the margin of its "feathers yellowiſh-brown, below whitiſh, the margin of its "equal tail-quills, and of its wing-quills, whitiſh."

The

The RUFOUS FLYCATCHER.

Le Gobe-Mouche Roux a Poitrine Orangée de Cayenne, Buff.
Muſcicapa Rufeſcens, Gmel.

Tenth Species.

The Rufous Flycatcher is found in Guiana in the ſkirts of the woods and the margins of the ſavannas: it is eaſily diſtinguiſhed, its breaſt being orange, and the reſt of its body rufous: its length is four inches nine lines; its bill is very flat and broad at the baſe; the head and the higher-part of the neck is greeniſh-brown; the back is rufous, ſtained alſo with greeniſh-brown; the tail is entirely rufous; the black of the wing-quills does not appear when they are cloſed except at the point, their ſmall webs being rufous: inſtead of the orange ſpot on the breaſt, white or whitiſh covers the under-part of the body. There is only one ſpecimen in the King's cabinet. [A]

[A] Specific character of the Rufous Flycatcher, *Muſcicapa Rufeſcens*, Gmel.: — " It is of a gloſſy tawny, below white, its " wing-quills black, a brown daſh on its top."

 The

The LEMON FLYCATCHER OF LOUISIANA, *Buff.*

Eleventh Species.

This Flycatcher may be compared for its ſize and colour to the Yellow Wagtail. Its breaſt and belly are covered with a fine lemon-colour, which is ſtill brighter on the forehead and the cheeks; the reſt of the head and neck are enveloped with a fine black, which extends below the bill, and forms a round horſe-ſhoe on the breaſt; a greeniſh-gray covers, on the back and ſhoulders, the cinereous ground of the plumage, and appears in lines on the ſmall webs of the great wing-quills. The vivacity and elegance of its colours, the gloſſy black conſpicuous on its light yellow ground, the uniform tint of its greeniſh robe, conſpire to render this bird one of the handſomeſt of the genus.

The RED-EYED FLYCATCHER.

Le Gobe Mouche Olive de la Caroline & de la Jamaique, Buff.
Muſcicapa Olivacea, Linn. and Gmel.
Muſcicapa Jamaicenſis, Briſſ.
The Olive-coloured Flycatcher, Edw.

Twelfth Species.

This conſiſts of two kinds of varieties; the one deſcribed by Edwards, the other by Cateſby. The

The firſt is of the ſame bulk and proportions as thoſe of the European Flycatchers. The upper-part of the head and body is olive-brown; a white fillet riſes above the eyes; the ground-colour of the quills is aſh-brown, and they are fringed with olive for a conſiderable part of their length.—The ſecond kind is deſcribed by Cateſby under the name of the *Red-eyed Fly-catcher;* its colours are darker than thoſe of the former. It breeds in Carolina, and re-moves to Jamaica in winter; but Sir Hans Sloane makes no mention of it. Brown how-ever reckons it one of the migratory Jamaica birds. It has not a great extent of notes, he tells us, but its tones are full and mellow. —This property muſt be peculiar to it, for all the other Flycatchers utter ſhrill broken ſounds. [A]

[A] Specific character of the Red-eyed Flycatcher, *Muſcicapa Olivacea*, LINN.: — "It is olive, below whiter, its eye-brows "white, its eyes red." In Jamaica it is called, on account of its note, *Whip Tom Kelly*. It makes a pendulous neſt, formed with wool and cotton, lined with hair and withered graſs, and bound together by a thready moſs. It lays five eggs, white, and thinly ſtrewed with rufous ſpots.

 The

The MARTINICO FLYCATCHER.

Le Gobe-Mouche Huppé de la Martinique, Buff.
Muſcicapa Martinica, Linn. and Gmel.
Muſcicapa Martinicana Criſtata, Briſſ.

Thirteenth Species.

A fine brown, which is deeper on the tail, covers all the upper-part of the body of this bird as far as the head, whoſe ſmall feathers, tinged with ſome ſtreaks of a more vivid rufous-brown, are half erect, forming a tuft on the crown: under the bill is a little white, which ſoon gives place to a light ſlate-gray, that covers the fore-part of the neck, the breaſt, and the ſtomach; the ſame white appears again on the belly. The quills of the wings are blackiſh-brown, fringed with white; their coverts, which are fringed with the ſame, enter by degrees into the rufous tint of the ſhoulders; the tail is ſomewhat tapered, its third-part hid by the wings, and is two inches long. The bird is five inches and a half. [A]

[A] Specific character of the *Muſcicapa Martinica*: — "Its head is creſted, its body brown, below cinereous, the exterior margin of its wing-quills whitiſh."

The

The BLACK-CAP FLYCATCHER.

Le Gobe-Mouche Noirâtre de la Caroline, Buff.
Muſcicapa Fuſca, Gmel.
Muſcicapa Carolinenſis Fuſca, Briſſ.
The Black-headed Flycatcher, Penn.

Fourteenth Species.

This bird is nearly as large as the Nightingale; its plumage, from the head to the tail, is of an uniform dull brown; its breaſt and belly are white, with a ſhade of yellowiſh-green; its thighs and legs are black; the head of the male is of a deeper black than that of the female, and this is the only difference between them. They breed in Carolina according to Cateſby, and migrate from thence on the approach of winter. [A]

[A] Specific character of the *Muſcicapa Fuſca:* — "It is "brown, below ochry-white; its bill, its top, and its legs "black."

The BLACK and WHITE FLYCATCHER.

Le Gillit, ou *Gobe-Mouce Pie de Cayenne,* Briſſ.
Muſcicapa Bicolor, Gmel.

Fifteenth Species.

This bird, which is called *Gillit* in its native country Guiana, is of an uniform white on the

 head,

head, the throat, and all the under-part of the body. The rump, the tail, and the wings, are black, and the ſmall quills of theſe edged with white. A black ſpot riſes behind the head, and ſtretches to the neck, where it is bounded by a white cap, which makes a circle on the back.—The length is four inches and a half, and the plumage of the female is entirely of a light uniform gray. It is found in the overflowed ſavannas.

The *White Bellied Flycatcher of Cayenne*, N° 566. fig. 3. *Pl. Enl.* hardly differs at all from the *Gillit*, and we ſhall not ſeparate them.

We ſhall alſo claſs with it the *White and Black Flycatcher* of Edwards, from Surinam, of which the colours are the ſame, except the brown on the wings, and black on the crown of the head, differences which are not ſpecific.

The CINEREOUS FLYCATCHER.

Le Gobe-Mouche Brun de la Caroline, Buff.
Muſcicapa Virens, Linn. and Gmel.
Muſcicapa Carolinenſis Cinera, Briſſ.

Sixteenth Species.

This is called by Cateſby, *The Little Brown Flycatcher*. Its figure and ſize are the ſame as thoſe

thoſe of his Olive Flycatcher with red eyes and legs, and we ſhould have ranged them together, had not that accurate obſerver diſtinguiſhed them. A dull brown tint, which covers uniformly all the upper-part, is interſected by the ruſty-brown of the feathers of the wings and tail; the under-part of the body is dirty white, with a ſhade of yellow; the thighs and legs are black; the bill is flat, broad, and a little hooked at the point, and eight lines in length; the tail is two inches; the whole length of the bird five inches eight lines; it weighs only three gros.—This is all that Cateſby informs us; and from him the reſt have borrowed their deſcriptions. [A]

[A] Specific character of the *Muſcicapa Virens*: — "It is "greeniſh-brown, below yellow, its eye-brows white."

The ACTIVE FLYCATCHER.

Le Gobe-Mouche de Cayenne, Buff.
Muſcicapa Agilis, Gmel.

Seventeenth Species.

This Flycatcher is not larger than the Yellow Wren of Europe; its plumage is almoſt the ſame, being cinereous and dirty white in both, only this little bird has a greater mixture of greeniſh. The flatneſs of its bill indicates its relation to the Flycatchers. Our Wrens however have the ſame inſtincts, and feed upon

 the

the various ſorts of flies: in ſummer, they continually circle in ſearch of the winged inſects; and in winter they attack their chryſalids and pierce the horny ſhell.

The total length four inches and a half; the bill ſeven lines; the tail twenty lines, and projects fifteen lines beyond the wings. [A]

[A] Specific character of the *Muſcicapa Agilis*. — "It is olive-"brown, below partly whitiſh; the quills of its wings and tail "black, and olive-brown at their margin."

The STREAKED FLYCATCHER.

Le Gobe Mouche Tacheté de Cayenne, Buff.
Muſcicapa Variegata, Gmel.

Eighteenth Species.

This Streaked Flycatcher is nearly of the ſame ſize as the Active Flycatcher, which is alſo a native of Cayenne. Dirty white, with a caſt of greeniſh on the wing, and ſome diſtincter ſpots of yellowiſh white, with aſh-brown on the head and neck, and blackiſh cinereous on the wings, form the confuſed mottled plumage of this bird. It has a ſmall beard of whitiſh briſtled feathers under the bill, and a half creſt of aſh-coloured feathers mixed with yellow filaments on the crown of the head. The bill is of the ſame ſize as that of the preceding, and the tail

is

is of the ſame length, but differs in its colours. The Active Flycatcher appears alſo more nicely formed, and more lively in its motions than the Streaked; at leaſt as far as we can judge from the ſtuffed ſpecimens.

The LITTLE BLACK AURORA FLY-CATCHER OF AMERICA.

Muſcicapa-Ruticilla, Linn. and Gmel.
Muſcicapa Americana, Briſſ.
The Black-headed Warbler, Lath. and Penn.
The Small American Redſtart, Edw.
The Small Black and Orange-coloured Bird, Ray.

Nineteenth Species.

We thus mark the two conſpicuous colours of the plumage of this bird, to which naturaliſts have hitherto given only the vague appellation of American Flycatcher. It is hardly ſo large as the Yellow Wren. A bright black is ſpread over the head, the throat, the back, and the coverts; a beautiful yellow aurora is pencilled on the white gray of the ſtomach, and deepens under the wings; it alſo appears in ſtreaks between the quills of the wings, and covers two-thirds of thoſe of the tail, both which are tipped with black, or blackiſh.—Such are the colours of the male. In the female the black is dilute blackiſh, and the orange and bluſh-colour yellow. Edwards gives figures of both male and female. Cateſby

Catesby represents the bird also under the name of *Small American Red-Start;* but it is rather of a larger size, which would make us presume that it is a variety. [A]

[A] Specific character of the *Muscicapa-Ruticilla:* — "It is "black; its breast, a spot on its wings, and at the base of the tail-"quills, yellow."

The ROUND-CRESTED FLYCATCHER.

Le Rubin, ou *Gobe-Mouche Rouge Huppé de la Riviere des Amazones,* Buff.
Muscicapa Coronata, Gmel.

Twentieth Species.

Of all the numerous family of Flycatchers this is the most brilliant. Its slender delicate shape suits the lustre of its garb: a crest, consisting of small divided feathers of fine crimson, projects in rays on its head; the same colour appears under its bill, covers its throat, breast, belly, and reaches the coverts of its tail; an ash-brown, intersected by some whitish waves on the edge of the coverts, and even of the quills, covers all the upper-part of the body and wings; the bill is very flat, and seven lines long; the tail two inches, and exceeds the wings by ten lines; the whole length of the bird is five inches and a half. Commerson calls it *Cardinal Titmouse*, though it is neither a Cardinal nor a Titmouse.

Titmouſe*. It would be one of the handſomeſt birds for the cage; but the nature of its food ſeems to place it beyond the dominion of man, and to enſure it liberty or death. [A]

* We found the figure of theſe birds among the drawings brought by Commerſon from the country of the Amazons. In Spaniſh it is called *Putillas*, as appears from a note at the bottom of the figure. The female, which is repreſented with the male, has no creſt; all the beautiful tints of its plumage are fainter.

[A] Specific character of the *Muſcicapa Coronata*: — "It is "brown; the creſt on its head roundiſh; its temples, and the "under-ſide of its body, red."

The RUFOUS FLYCATCHER.

Le Gobe-Mouche de Cayenne, Buff.
Muſcicapa Rufeſcens, Gmel.

The Twenty-firſt Species.

This Flycatcher, which is five inches and a half long, is nearly of the ſize of the Nightingale; all the upper-part of its body is of a fine light rufous, with a flame caſt, which extends over the ſmall quills of the wings, and theſe covering the great quills when the wings are cloſed, have only a ſmall black triangle formed by their extremities; a brown ſpot covers the crown of the head; all the anterior and the upper-parts of the body, are tipt with ſome ſlight ſhades of rufous; the tail is ſquare and ſpread; the bill is broad, ſhort, and ſtrong, and its point reflected, and

partaking

partaking therefore both of the Flycatchers and of the Tyrants. We are uncertain whether to refer it to Briſſon's Rufous Flycatcher of Cayenne.—It is a diſcouraging circumſtance that nomenclators have ſo often claſſed diſtinct objects by the ſame name: however, the *Rufous Flycatcher of Cayenne*, is, according to Briſſon, eight inches long, and ours is only five; and the difference in regard to colour will appear from comparing his deſcription with what we have given *. But there is no eſſential diſtinction between them, except in regard to ſize; and that difference cannot be imputed to age, for if the ſmaller were ſuppoſed to be the younger, the orange ſpot on the breaſt would be leſs vivid than in the adult. [A]

* "Above, tawny rufous; below, dilute rufous; its head, throat, and neck, deep cinereous; the feathers on its throat, and its lower-neck, edged with whitiſh; its breaſt, rump, and tail-quills, bright rufous." BRISSON.

[A] Specific character of the *Muſcicapa Rufeſcens*:—"It is of "a gloſſy tawny; below white; its tail-quills black; a brown "daſh on its top."

The YELLOW-BELLIED FLYCATCHER.

Le Gobe-Mouche à Ventre Jaune, Buff.
Muſcicapa Cayenenſis, Linn. Gmel. and Briſſ.

Twenty-ſecond Species.

This beautiful Flycatcher inhabits the continent of America, and the adjacent iſlands. The one

one figured in the *Planches Enluminées* was brought from Cayenne; we have received another from St. Domingo, under the name of *Crested Flycatcher of St. Domingo*. We are of opinion that these differ only by their sex: that of St. Domingo seems to be the male; for the golden yellow of its crown is more vivid and more spread than in the other, where the lighter tint scarce appears through the blackish feathers which cover that part of the head. In other respects the two birds are similar. They are rather smaller than the Nightingale, being five inches and eight lines long; the bill is eight lines, and scarcely curved at the tip, and the wings reach not to the middle; the orange spot on the head is edged with a blackish ash-colour, a white bar crosses the face over the eyes, below which a spot of the same colour appears that spreads, and is lost in the rusty-brown of the back; this rusty-brown covers the wings and the tail, and becomes rather more dilute on the edge of the small webs of the quills; a fine orange-yellow covers the breast and the belly, which vivid colour distinguishes this bird from all the other Flycatchers. Though the golden yellow feathers of the crown can be erected at pleasure, as in the small European Wrens, yet, since they are usually reclined, the bird is not properly *a Crested Flycatcher*. [A]

[A] Specific character of the *Muscicapa Cayenensis*:—"It is "brown; below yellow; its eyebrows white; its top somewhat "orange."

The

The KING OF THE FLYCATCHERS.

Le Roi des Gobe-Mouches, Buff.

Twenty-third Species.

This has been named *The King of the Flycatchers*, on account of a beautiful crown placed tranſverſely on its head; whereas in all other birds the creſts lie longitudinally. It conſiſts of four or five rows of ſmall round feathers, ſpread like a fan, ten lines broad, all of a bright bay colour, and terminated with a little black ſpangle; ſo that it might be taken for a peacock's tail in miniature.

This bird is alſo remarkably ſhaped, and ſeems to combine the features of the Flycatchers, of the *Moucherolles*, and of the Tyrants. It is ſcarcely larger than the European Flycatcher, and has a diſproportioned bill, which is ten lines in length, and very broad and flat, beſet with briſtles that reach almoſt to its tip, which is hooked. The *tarſus* is ſhort; the toes ſlender; the wing is not more than three inches, nor the tail more than two. It has a ſmall white eye-brow; its throat is yellow; a blackiſh collar encircles its neck, and joins that tinge which covers the back, and changes on the wing into a deep fulvous brown. The quills of the tail are light bay; and the ſame colour, though more dilute, ſtains the rump and the belly; the whitiſh colour of the ſtomach is croſſed by ſmall blackiſh waves.—This bird is very rare; only one ſpecimen has been brought from Cayenne, where even it ſeldom appears.

The DWARFISH FLYCATCHERS.

Les Gobe-Moucherons, Buff.

Twenty-fourth and Twenty-fifth Species.

Nature has proportioned theſe birds to their feeble prey; a large American beetle might be a match for them.—We have ſpecimens of them in the King's cabinet, and a ſhort deſcription will ſuffice.

The firſt * is the ſmalleſt of the Flycatchers; it is leſs than the ſmalleſt of our Wrens, and in its figure, and even in its colours, it is nearly the ſame. Its plumage is olive, without any yellow on the head, but a few light ſhades of greeniſh appear on the lower-part of its back and on its belly; and ſmall lines of yellowiſh white are traced on the blackiſh quills, and on the coverts of the wings.—It is found in the warm parts of America.

The ſecond † is ſtill ſmaller than the firſt; all the under-part of its body is light yellow, verg-

* This is the *Muſcicapa Pygmæa* of Gmelin, and the *Dwarf Flycatcher* of Latham.

† This is the *Muſcicapa Minuta* of Gmelin, and the *Petty Flycatcher* of Latham.

ing on ſtraw colour; it is hardly three inches long; its head, and the beginning of the neck, are partly yellow, partly black, each yellow feather having in its middle a black ſtreak, which ſhews the two colours diſpoſed in long and alternate ſpots; the feathers of the back, the wings, and their coverts, are black cinereous, and edged with greeniſh; the tail is very ſhort, the wing ſtill ſhorter; the bill is ſlender, and lengthened, which gives this little Flycatcher a peculiar appearance.

The uſeful deſtination of the Flycatchers will occur to the moſt ſuperficial obſerver. The inſect tribes elude the interference of man; and though deſpicable as individuals, they often become formidable by their numbers. Inſtances are recorded of their multiplying to ſuch an amazing degree as to darken the air; of their devouring the whole vegetable productions; and of their carrying in their train the accumulated ills of famine and peſtilence. Happily for mankind ſuch calamities are rare, and Nature has wiſely provided the proper remedies. Moſt birds ſearch for inſects' eggs; many feed on their groveling *larvæ*; ſome live upon their cruſtaceous cryſalids; and the Flycatchers ſeize them after they eſcape from priſon, exulting on their wings. Hence in autumn, when theſe birds migrate into other climates, the ſwarms of gnats,

gnats, flies, and beetles, are in our latitudes more than uſually numerous. But in the tropical countries, where heat and moiſture conſpire to ripen the exuberance of inſect life, the Flycatchers are more eſſential. All Nature is balanced, and the circle of generation and deſtruction is perpetual! The philoſopher contemplates with tender melancholy this cruel ſyſtem of war; he ſtrives in vain to reconcile it with his ideas of benevolence of intention: but he is forcibly ſtruck with the nice adjuſtment of the various parts, their mutual connexion and ſubordination, and the unity of plan which pervades the whole.

The MOUCHEROLLES.

WE ſhall term thoſe Moucherolles which are larger than the common Flycatchers, but ſmaller than the Tyrants; and to avoid confuſion, we ſhall range them in two diviſions, correſponding to their ſize. As the Moucherolles are intermediate between the Flycatchers and the Tyrants, they participate of the nature of both.

They are found in both continents; but they are different ſpecies which occur in each. The ocean that intervenes between the tropics is the great barrier, which none but the palmipede birds, from their facility in reſting on the water, can paſs.

In the hot climates Nature ſports in the luxuriance of her productions. Many ſpecies of birds, ſuch as the Widow-birds, the Moucherolles, and the Bee-eaters, which inhabit thoſe ſultry regions, are furniſhed with tails of uncommon length: this character diſtinguiſhes the Moucherolles from the Flycatchers, from which they differ alſo in having their bill ſomewhat ſtronger, and more hooked at the tip.

The

No. 113

THE FORKED-TAIL FLYCATCHER.

The SAVANA.

Le Savana, Buff.
Muscicapa-Tyrannus, Linn. and Gmel.
The Fork-tail Flycatcher, Penn. and Lath.

First Species.

This Moucherolle is nearly as large as the Tyrants, and is figured in the *Planche Enluminées*, under the appellation of *Forked-tail Tyrant of Cayenne*; it is distinguished however by its bill, which is more slender and not so much hooked as in the Tyrants. It is called *the Widow* at Cayenne; but this name is appropriated to another kind of birds, which it resembles in nothing except the length of its tail. It constantly haunts the flooded savannas, and for that reason we have termed it the *Savana*. It is observed to perch upon the adjacent trees, and to alight every minute upon the clods or grassy tufts which rise above the surface of the water, jerking its tail like the Wagtails. It is as large as the Crested Lark; the quills of its tail are black, the two outermost nine inches long, and forked, the two following only three inches and a half, and the rest gradually shorter, so that the two mid-ones are only an inch:—and thus, though the bird is fourteen inches long, measuring from the point of the bill to the end of the tail, the distance between its bill and its nails is is only ſix inches. On the crown of its head is a yellow ſpot, which is however wanting in many ſubjects, theſe being probably females. A ſhort blackiſh ſquare hood covers the back of its head; beyond that, the plumage is white, which colour advances under the bill, and ſpreads over all the anterior and under-part of the body; the back is greeniſh-gray, and the wing brown. —This bird is found on the banks of the river De la Plata, and in the woods of *Montevideo*, from whence it was brought by Commerſon. [A]

[A] Specific character of the *Muſcicapa-Tyrannus*: — " Its " tail is very long and forked, its body black below white." It is found as far north as Canada.

The CRESTED MOUCHEROLLE, *with Steel-coloured Head.*

Muſcicapa-Paradiſi, Linn. and Gmel.
The Pied Bird of Paradiſe, Edw.
The Paradiſe Flycatcher, Lath.

Second Species.

This bird is found at the Cape of Good Hope, Senegal, and Madagaſcar. Briſſon deſcribes it in three different places of his ornithology, by the names of *the Creſted Flycatcher of the Cape of Good Hope* *, *the White Flycatcher of the Cape* of

* " Creſted Flycatcher:— "Above dilute ſcarlet, below white; " the breaſt cinereous white; the head and the upper-part of the " neck greeniſh-black; the tail-quills dilute purple."

of Good Hope *, and *the Crested Flycatcher of Brazil* †. These three are really the same, the first and third being males, and the second, which is rather larger, a female; a property which, though principally confined to the birds of prey, obtains also in the Flycatcher, the Moucherolles, and the Tyrants.

The male is seven inches long, the female eight inches and one-fourth; this excess being almost entirely in the tail; but its body is also somewhat thicker, and of the size of a common Lark: in both, the head and the top of the neck are covered, as far as the circular division in the middle, with black, shining with a green or bluish gloss, whose lustre is like that of burnished steel: its head is decorated with a beautiful crest, which falls loosely back; its eyes are flame-coloured; its bill is ten lines in length, a little arched near the tip, reddish, and beset with pretty long bristles. All the rest of the body of the female is white, except the great quills, through which the black appears at the tips of the wings when closed; there are two rows of black streaks on the small quill-feathers and in the great coverts; and the shafts of the tail-quills are uniformly black throughout.

* White Crested Flycatcher:—"The head and upper-part of "the neck greenish-black; the tail-quills white, their outer edges "and shafts black."

† Crested Flycatcher:—"Above dilute scarlet; below white; "the head greenish-black; the superior coverts of the wings gold-"coloured, the tail-quills dilute scarlet."

 In

In the male, the breaſt, below the black hood, is bluiſh-gray, and the ſtomach and all the under-part of the body white: a bright bay robe covers all the upper-part to the end of the tail, which is oval ſhaped and regularly tapered, the two middle quills being the largeſt, and the others ſhortening two or three lines each: the ſame is the caſe in the female.

According to Adanſon *, this Moucherolle lodges among the mangrove-trees, which grow in the ſolitary and unfrequented ſpots along the banks of the Niger and of the Gambra. Seba places it in Brazil, and ranges it with the birds of Paradiſe, applying the Brazilian appellation *Acamacu* †; but little can be relied upon the accuracy of that collector of Natural Hiſtory, who ſo often beſtows names without diſcernment. It is very unlikely that this bird could be found both in Africa and Brazil; yet Briſſon founds his claſſification upon the authority of Seba, at the ſame time that he expreſſes a ſuſpicion, that Seba was miſtaken. Klein ſuppoſes it to be a *Creſted Thruſh* ‡, and Moehring a Jackdaw ‖;—a ſtriking inſtance of the confuſion bred by a rage for nomenclature. But we have ſtill another: Linnæus imagines it to be a Raven; but as it has a long tail, he calls it the *Paradiſe Raven* §. [A]

* Supplement de l'Encyclopedie, tome i.

† Braſilian Paradiſe-bird, or Creſted Cuiriri Acamacu.

‡ Turdus Criſtatus. ‖ Monedula. § Corvus Paradiſi.

[A] Specific character of the *Muſcicapa Paradiſi*: — "Its "head creſted and black; its body white; its tail wedge-ſhaped; "its intermediate tail quills longeſt."

The

The VIRGINIAN MOUCHEROLLE.

Muſcicapa Carolinenſis, Linn. and Gmel.
The Cat Flycatcher, Penn. and Lath.

Third Species.

Cateſby calls this the *Cat-bird*, becauſe its cry reſembles the mewing of a cat. It paſſes the ſummer in Virginia, where it feeds upon inſects; it does not perch on large trees, and frequents only the ſhrubs and buſhes. *It is a little larger*, he tells us, *than a Lark.* Its ſize is therefore nearly the ſame as that of the Little Tyrant; but the ſtraightneſs of its bill diſtinguiſhes it from the Tyrants. The plumage is dark, being variouſly mixed with black and brown: the upper-ſide of its head is black, and the upper-ſide of its body, of its wings, and of its tail, deep brown; an even blackiſh on the tail: its neck, its breaſt, and its belly are of a lighter brown; a dull red caſt appears on the lower coverts of its tail, which is three inches long, and conſiſts of twelve equal quills, and only two-thirds of it covered by the wings; the bill is ten lines and a half, and the whole length of the bird is eight inches. —It breeds in Virginia, and lays blue eggs; it migrates on the approach of winter. [A]

[A] Specific character of the Cat Flycatcher, *Muſcicapa Carolinenſis*, LINN.: — "It is brown, below cinereous, its head "black, its vent red." It builds its neſt with leaves and ruſhes, and lines it with fibrous roots. It is very courageous, and will attack a crow.

The BROWN MOUCHEROLLE of MARTINICO.

Muscicapa Martinica, Gmel.
Muscicapa Martinicana Cristata, Briss.

Fourth Species.

This Moucherolle has not a long tail like the preceding kinds; in its size and figure it resembles the largest of the Flycatchers. It is distinguished from the Tyrants by the shape of its bill, which is not so much hooked as the bill of the smallest Tyrants, and more slender; it is however eight lines long, and the bird itself six inches and a half. A deep brown of a pretty uniform tinge covers the upper-part of the body, the head, the wings, and the tail; the under surface of the body is undulated with transverse waves of rufous brown; a few reddish feathers form the inferior coverts of the tail, which is square, and the edges of its outer-quills are fringed with white lines. [A]

[A] Specific character of the *Muscicapa Martinica*: — "Its "head is crested, its body brown, below cinereous, the outer "margin of its wing-quills whitish."

The

The FORKED-TAIL MOUCHEROLLE of MEXICO, *Buff.*

Muſcicapa Forficata, Gmel.
The Swallow-tailed Flycatcher, Lath.

Fifth Species.

It is larger than the Lark; its whole length is ten inches, of which its tail meaſures five; its eyes are red, its bill eight lines long, flat, and rather ſlender: its head and back are covered with a very light gray, mixed with a dilute reddiſh; the red colour below the wings extends alſo on the ſides, and tinges the white that is ſpread over the whole of the under-ſide of the body; the ſmall coverts are aſh-coloured, and edged with ſcaly white lines; the great coverts, which are blackiſh, are ſimilarly fringed; the great quills of the wings are entirely black, and ſurrounded with ruſty-gray: the outermoſt quills of the tail are the longeſt, and are forked like the Swallow's tail: the other quills diverge leſs, and gradually ſhorten; ſo that the middle one is only two inches long: they are all of a gloſſy black, and fringed with ruſty-gray: the outer webs of the largeſt quills on each ſide appear white almoſt their whole length. Some ſpecimens have the tail longer than that ſent from Mexico by M. de Boynes, then Secretary for the Marine Department.

The

The MOUCHEROLLE of the PHILIPPINES.

Muscicapa Philippensis, Gmel.

Sixth Species.

It is as large as the Nightingale; all the upper-part of its body is brown-gray; all the under-part of the wings and tail are whitish from below the bill; a white line stretches over the eyes, and long diverging hairs appear at the corners of the bill. Such are the obscure ambiguous features of this bird. A specimen is lodged in the King's Cabinet.

The GREEN-CRESTED VIRGINIAN MOUCHEROLLE, *Buff.*

Muscicapa Crinita, Linn and Gmel.
The Crested Flycatcher, Penn. Cat. and Lath.

Seventh Species.

The length of the tail and bill of this bird marks its relation to the Moucherolles: it is rather larger than the Flycatchers, being eight inches long, of which its tail forms the half; its bill is flat, beset with bristles, and scarcely hooked at the tip, and it measures twelve lines and a half; the head is furnished with small

ſmall feathers reclined into a half-creſt; the top of the neck, and all the back, dull green; the breaſt and the fore-part of the neck leaden-gray; the belly of a fine yellow; the wings brown, and ſo are the great quills which are edged with bay; thoſe of the tail are the ſame. This bird is not ſhaped like the Tyrants, but appears to partake of their gloomy ſullen temper. It would ſeem, ſays Cateſby, from its diſagreeable ſcreams, to be always in enmity, and continually at variance with the other birds. It breeds in Carolina and Virginia, and before winter it removes to hotter climates. [A]

[A] Specific character of the Creſted Flycatcher, *Muſcicapa Crinita*: — "Its head is creſted, its neck bluiſh, its belly yellowiſh, its back greeniſh, and the quills of its wings and tail "rufous." It builds its neſt in the holes of trees, employing for the materials, hair and ſnakes ſkins.

The SCHET of MADAGASCAR.

Muſcicapa Mutata, Linn. and Gmel.
The Mutable Flycatcher, Lath.

Eighth Species.

The name *Schet* is applied in Madagaſcar to a beautiful long-tailed *Moucherolle;* and two others are called *Schet-all*, and *Schet-Vouloulou*, which ſeem to denote the Rufous Schet and the Variegated Schet, and mark only two varieties of the ſame ſpecies. Briſſon reckons three; but a few differences

differences in the colours are not ſufficient to conſtitute diſtinct ſpecies, where the ſhape, the ſize, and all the other proportions, are the ſame.

The Schets have the long form of the Wagtail; they are rather larger, meaſuring ſix inches and a half to the end of the true tail, not to mention two feathers which extend almoſt five inches farther; the bill is ſeven lines, triangular, very flat, broad at the baſe, beſet with briſtles at the corners, and with hardly any perceptible curve at the point: a beautiful blackiſh-green creſt, with the luſtre of burniſhed ſteel, is bent ſmooth back, and covers the head; the iris is yellow, and the eye-lid blue.

In the firſt variety, the ſame dark colour that paints the creſt, encircles the neck, and inveſts the back, the great quills of the wings and of the tail, of which the two long feathers meaſure ſeven inches, and are white, as are alſo the ſmall quills of the wings, and all the under-part of the body.

In the *Schet-all*, the colour of the creſt appears only on the great quills of the wings, whoſe coverts are marked with broad white lines; all the reſt of the plumage is a bright gilded bay, which Edwards terms a *fire ſhining cinnamon*, which is ſpread equally ove the tail and the two long projecting ſhafts; theſe ſhafts are ſimilar to thoſe which are ſent of from the tail in the Angola and Abyſſinian Rdlers, only in in these birds they are the outermost, while in the Madagascar Moucherolle they occupy the middle.

The third variety, or the *Schet-Vouloulou*, has scarcely any difference from the preceding, except that the two projecting feathers of the tail are whitish; the rest of the plumage is bay-coloured, as in the *Schet-all*.

In the *Schet-all* which is preserved in the King's Cabinet, these two feathers are six inches long; in another specimen, I found them to be eight inches, and the outer webs edged with black three-fourths of their length, and the remainder white; in a third, these two long feathers were entirely wanting; whether we must impute this to some accident, to the age, or to the moulting, which Edwards thinks lasts six months in these birds *?

They are found not only in Madagascar, but in Ceylon, and at the Cape of Good Hope. Knox gives a good description of them †. Edwards

* "I received this bird (the *Schet-all*) from Ceylon. M. Brisson says, that it comes from the Cape of Good Hope; but the figure which he gives of it is surely imperfect, as it has not the two feathers of the tail, which are so remarkably large. I believe it is natural to some birds which have these long tails, to want them six months in the year . . . . which I have seen in some long-tailed birds at London. . . . The White Crested Flycatcher described by Brisson, is certainly the male of the same species."

GLEANINGS.

† "They are small birds, not much exceeding Sparrows, charming to the eye, but good for nothing else. Some of these birds have their bodies as white as snow, the quills of their tail a

foos

Edwards calls the third Schet-all *the Pied Bird of Paradiſe*; however, Schets are totally different from the Birds of Paradiſe.

foot long, and their heads black like jet, with a tuft or creſt. There are many others of the ſame kind, the only difference conſiſting in the colour, which is reddiſh-orange: theſe birds have alſo a tuft of black feathers erect on the head. I believe the one ſort are the males, and the others the females of the ſame ſpecies."

Hiſt. of Ceylon, by Robert Knox, *London*, 1681.

The TYRANTS.

THE appellation of Tyrant applied to these birds muſt appear whimſical. According to Belon, the ancients termed the Little Crowned Wren *Tyrannus* *: in the preſent caſe, the name refers not only to this crown, but alſo to their ſanguinary diſpoſition. A ſad proof of human miſery, that the idea of cruelty is ever conjoined with the emblem of power! We ſhould therefore have changed this mortifying and abſurd term, but we found it too firmly eſtabliſhed by naturaliſts :—It is not the firſt time that we have been compelled by the general uſage to acquieſce in improper and incongruous epithets.

Theſe inhabitants of the New World are larger than the Flycatchers or Moucherolles; they are ſtronger and more vicious; their bill is larger and firmer; their diſpoſitions are darker and more audacious, and, in this reſpect, they reſemble the Shrikes, to which they are analogous alſo in the ſize of their body and the ſhape of their bill.

* This word, in Greek, ſignifies merely a king or prince.

The TITIRI, or PIPIRI.

Lanius-Tyrannus, Linn. Gmel. and Borouſk.
Muſcicapa Tyrannus, Briſſ.
Pica Americana Criſtata, Friſch.
Turdus Coronâ Rubrâ, Klein.
The Tyrant Shrike, Lath.

The Firſt and Second Species.

IT has the ſize and ſtrength of the Great Cinereous Shrike; it is eight inches long, thirteen inches of alar extent; its bill flat, but thick, and thirteen lines long, briſtled with muſtachoes, and ſtraight to the tip, where it is hooked: its tongue is acute and cartilaginous; the feathers on the crown of its head are yellow at the root, and terminated with a blackiſh ſpeckling, which covers the reſt when they are flat, but, when the bird ſwells with rage, they become erect, and the head then appears crowned with a broad tuft of the moſt beautiful yellow; a light brown-gray covers the back, and on the ſides of the neck it melts to the white ſlate-gray of the anterior and under-part of the body: the brown quills of the wing and of the tail are edged with a ruſty thread.

The female has the yellow ſpot on the head though not ſo broad, and its colours are more dilute, or duller than thoſe of the male. A female, meaſured at St. Domingo by the Chevalier

N.o 114

THE GRAND TYRANT.

Deſhayes, was an inch longer than the male, and its other dimenſions in proportion: hence, in general, the ſmalleſt individuals in this ſpecies are the males *.

At Cayenne, this Tyrant is called *Titiri*, from the reſemblance to its ſhrill noiſy ſcreams. The male and female keep commonly together in the cleared ſpots of the foreſts; they perch on the lofty trees; and are very numerous in Guiana. They breed in the hollow trunks, or in the clefts of the branches below the ſhade of the moſt leafy bough. If one attempts to plunder their young, their natural audacity changes into intrepid fury; they contend obſtinately; they dart upon the perſon; purſue him; and if, in ſpite of all their exertions, they are unable to reſcue their dear offspring, they fondly viſit the cage, and carry food.

This bird, though ſmall, appears to dread no ſort of animal. "Inſtead of fleeing, like the other birds," ſays Deſhayes, "or concealing itſelf from the rapacious tribes, it attacks them with intrepidity, and haraſſes them to ſuch a degree, that it generally ſucceeds in driving them off. No animal dares to come near the

* "All the Pipiris are not exactly of the ſame ſize or of the ſame plumage; beſides the difference remarked in all the kinds between the male and the female, there is ſtill another with reſpect to the bulk of individuals in this ſpecies. This difference is often perceived, and ſtrikes even the moſt careleſs obſervers. Probably the abundance or ſcarcity of proper food is the cauſe of the diverſity." *Note communicated by the* CHEVALIER DESHAYES.

tree where it breeds. It purſues to a conſiderable diſtance, and with implacable obſtinacy, all that it conceives to be its enemies, dogs eſpecially, and birds of prey *." It is not even intimidated at man; ſo lately has his empire been eſtabliſhed in thoſe ſavage countries, that it ſeems not conſcious of his power †. In the moments of its fury it ſhuts its bill forcibly, which occaſions a quick repeated cracking.

In St. Domingo this bird is named *Pipiri*, which, as well as *Titiri*, expreſſes its uſual cry or ſquall. It is diſtinguiſhed into two varieties, or two contiguous ſpecies: the firſt is the *Great Pipiri*, of which we have juſt ſpoken, and which is called in that country *The Black-headed Pipiri*, or *The Thick-billed Pipiri;* the other is called *The Yellow-headed Pipiri*, or *The Migratory Pipiri*, and is ſmaller and weaker. The upper-part of the body in the laſt is gray, fringed throughout with white; but in the Great Pipiri it is fringed with rufous. The diſpoſition of the ſmall Pipiris is alſo much milder, and not ſo ſavage as the others. Theſe remain ſequeſtered in the wilderneſs, and are never met with except in pairs; while the ſmall Pipiris appear often in troops, and come near the ſettlements.

* M. Deſhayes.

† "I ſhot a young one, which was only ſlightly wounded. My little negro who ran after it was attacked by a Shrike of the ſame ſpecies, which was probably the mother: this bird fixed with ſuch rancour on the boy's head, that he had the utmoſt difficulty to get rid of it." *Note communicated by M. de* Manoncour.

They

They aſſemble in conſiderable flocks during the month of Auguſt, and haunt thoſe places which yield certain kinds of berries that attract the beetles and inſects. At that time theſe birds are very fat, and are caught for the table*.

Though they are called Migratory Pipiris, it is not probable, ſays Deſhayes, that they ever quit the iſland of St. Domingo, which is of ſufficient extent to admit local changes. In fact, they leave their uſual haunts in certain ſeaſons, and follow the maturity of the fruits which feed their inſect prey. All their other habits are the ſame as thoſe of the Great Pipiris: both ſpecies are very numerous in St. Domingo, and few birds occur in more frequency†.

They live upon caterpillars, beetles, butterflies, and waſps. They perch on the higheſt ſummit of trees, and eſpecially on the palms, from thence they deſcry the inſect as it roves in the air, and the inſtant that they ſeize it they return again to their bough. They ſeem moſt engaged from ſeven in the morning till ten;

* M. Deſhayes.

† "They are ſeen in the foreſts, in the abandoned grounds, in the cultivated ſpots; they like every ſituation; yet the ſpecies of the Yellow-headed Pipiris, which are the moſt numerous, ſeem to prefer the ſettled parts. In winter they come near the houſes; and as this ſeaſon from the mildneſs of the climate correſponds to the ſpring in France, it ſeems that the coolneſs which then prevails inſpires them with cheerfulneſs. Never are they ſeen ſo noiſy, or ſo joyous, as in the months of November and December. They frolic with each other, toy, and careſs." *Note communicated by M.* Deſhayes.

and again from four o'clock in the afternoon till ſix. It is amuſing to ſee them hunting their fugacious prey, and purſuing their devious courſe; but their lofty conſpicuous ſtation expoſes them perpetually to the eye of the fowler.

No birds are ſo early awake as the Pipiris; they are heard at the firſt appearance of dawn*; they paſs the night on the ſummits of the talleſt trees, and hail the approach of the morning. There is no ſtated ſeaſon for their amours†. They breed, ſays M. Deſhayes, *in the heats of autumn, and during the freſhening air of winter*, at St. Domingo, though ſpring is the moſt uſual ſeaſon; they lay two or three eggs, ſometimes four, which are whitiſh, and ſpotted with brown. Barrere reckons this bird a Bee-eater, and terms it *Petit-ric*. [A]

* "Except the Cock, the Peacock, and the Nightingale, which ſing during the night, no bird is ſo early." *Note communicated by M.* Freſnaye, *formerly Counſellor at* Port-au-Prince.

† "The Black-headed Pipiris lay moſt undoubtedly in December. We cannot affirm whether each female breeds every year; nor whether theſe winter hatches, which ſeem extraordinary, be not occaſioned by accidents, and deſtined to repair the loſs of hatches made in the proper ſeaſon." *Note communicated by M.* Deſhayes.

[A] Specific character of the *Lanius-Tyrannus*:—"It is cinereous; below white; its top black; a longitudinal ſtreak, fulvous."

The

The TYRANT OF CAROLINA.

Lanius-Tyrannus, var. 3. Linn. and Gmel.

Third Species.

From the account which Catesby has given of this bird, we do not hesitate to class it with the Pipiri of St. Domingo, since its disposition and its habits are the same*. But it is distinguished by its red crown, and the manner of placing its nest, which is left entirely exposed in the shrubs or bushes; whereas the Pipiri conceals its nest, or even lodges it in the holes of trees. It is nearly of the same size as the Great Pipiri: its bill seems less hooked: Catesby says only *that it is broad, flat, and tapering.* The red

* "The courage of this little bird is singular. He pursues and puts to flight all kinds of birds that come near his station, from the smallest to the largest, none escaping his fury; nor did I ever see any that dared to oppose him while flying, for he does not offer to attack them when sitting. I have seen one of them fix on the back of an Eagle, and persecute him so that he has turned on his back into various postures in the air, in order to get rid of him; and at last was forced to alight on the top of the next tree, from whence he dared not to move till the little Tyrant was tired, or thought fit to leave him.—This is the constant practice of the cock, while the hen is breeding; he sits on the top of a bush, or small tree not far from her nest, near which if any small birds approach, he drives them away; but the great ones, as Crows, Hawks, Eagles, he will not suffer to come within a quarter of a mile of him without attacking them. They have only a chattering note, which they utter with great vehemence all the time they are fighting.—When their young are flown they are as peaceable as other birds." CATESBY.

 spot

ſpot on the upper-part of its head is very brilliant, and is encircled with black feathers, which conceal it when they are cloſed.—This bird appears in Virginia and Carolina about the month of April; there breeds, and departs in the beginning of winter.

A bird ſent to the King's cabinet, under the name of *Louiſiana Tyrant*, appears to be exactly the ſame with the Carolina Tyrant of Cateſby. It is larger than the fifth ſpecies, or Cayenne Tyrant, and almoſt equal to the Great Pipiri of St. Domingo. An aſh-colour, almoſt black, is ſpread over all the upper-part of the body, from the crown of the head to the end of the tail, which terminates in a ſmall white bar ſhaped into feſtoons; light whitiſh waves are intermixed in the ſmall quills of the wing; ſome ſmall ſtreaks of deep orange, inclined to red, ſhine through the blackiſh quills on the top of the head; the throat is of a pretty pure white, which is ſhaded with black on the breaſt, and again becomes ſnowy from the ſtomach as far as the tail. [A]

[A] The Carolina Tyrant builds its neſt with wool and moſs, and lines it with fibrous roots. It lays five eggs, which are white, with ruſty ſpots.

The

The BENTAVEO, *or* The CUIRIRI.

Lanius-Pitangua, Linn. and Gmel.
Pitangua-guacu, Ray and Will.
Tyrannus Brasiliensis, Briss.
The Brasilian Shrike, Lath.

Fourth Species.

This Tyrant, called *Bentaveo* at Buenos-Ayres, whence it was brought by Commerson, and *Pitangua-guacu* by the people of Brazil, has been described by Marcgrave*. He makes it of the size of the Stare (we will observe that it is thicker, and more bulky); and represents its bill as thick, broad, and pyramidal, its edges sharp, and more than an inch long; its head bulky; its neck short; the head, the top of its neck, the whole of its back, its wings, and its tail, of a blackish brown, slightly shaded with dull green;

* "The Pitangua-guacu of the Brazilians, Bemtere of the Portuguese, is equal in bulk to the Stare; has a bill thick, broad, pyramidal, somewhat more than an inch long, sharpened exteriorly; its head compressed, and broadish; its neck short, which it contracts when sitting. Its body is nearly two inches and a half long; its tail broadish, and three inches long; its legs and feet are brown. Its head, the upper-part of its neck, the whole of its back, its wings, and its tail, are of a blackish brown, mixed with a very little greenish. The lower-part of its neck, its breast, and its lower-belly, have yellow feathers; the upper-part, however, near the head, has a little crown of white. From below the throat to the origin of the bill is white. It calls with a loud voice. Some of these birds have a yellow spot on the top of the head; some have it partly yellow; they are called by the Brazilians, Cuiriri. In every other respect they are like the *Pitangua-Guacu.*" MARCGRAVE.

its

its throat white, and alſo the little bar on the eye; the breaſt and belly yellow; and the ſmall quills of the wings fringed with ruſty colour. Marcgrave adds, that ſome of theſe birds have an orange ſpot on the crown of the head, and others a yellow one. The Brazilians call theſe *Cuiriri;* and in every other property they are ſimilar to the *Pitangua-guacu.* Seba applies the name *Cuiriri* to a ſpecies entirely different.

Thus the Bentaveo of Buenos-Ayres and the Pitangua and Cuiriri of Brazil are the ſame; and in their inſtincts ſimilar to the Great Pipiri of St. Domingo, or the Titiri of Cayenne: but the colours of the Bentaveo, its bulk, and the thickneſs of its bill, the moſt obviouſly diſtinguiſh it. [A]

[A] Specific character of the *Lanius-Pitangua:*—"It is black; "below white; a yellow ſtreak on its top; a white belt on its "eyes."

The CAYENNE TYRANT.

Muſcicapa Ferox, Gmel.
Tyrannus Cayannenſis, Briſſ.
The Tyrant Flycatcher, Lath.

Fifth Species.

It is larger than the Red-backed Shrike of Europe. In the ſpecimen belonging to the King's cabinet all the upper-part of the body is aſh-gray, deepening into black on the wings, of which

which ſome quills have a light white border; the tail is of the ſame dark caſt, and is pretty broad, and three inches long; the whole bird meaſures ſeven inches, and the bill ten lines; a lighter gray covers the throat, and receives a greeniſh tinge on the breaſt; the bill is of a ſtraw, or light ſulphur colour; the ſmall feathers on the top, and anterior part of the head, are half erect, and are painted with ſome ſtrokes of citron-yellow and aurora-yellow; the bill is flat, beſet with briſtles, and hooked at the point. The female is not of ſo deep a brown.

The Little Cayenne Tyrant of the *Planches Enluminées* is rather ſmaller than the preceding, and only a variety of it. The one deſcribed by Briſſon is alſo a variety. [A]

[A] Specific character of the *Muſcicapa Ferox*:—"It is brown; "its chin, its throat, and its breaſt, cinereous; its belly yellowiſh; "the greater quills of its wings olive at their margin."

The CAUDEC.

Muſcicapa Audax, Gmel.
The Yellow-crowned Flycatcher, Lath.

Sixth Species.

This is the *Spotted Flycatcher of Cayenne*, as repreſented in the *Planches Enluminées*; but the hooked form of its bill, its ſtrength, its ſize, and its diſpoſition, entitle it to the name of Tyrant.

It

It is called *Caudec* at Cayenne, and is eight inches long; the bill is ſcalloped at the edges near the hooked point, is beſet with briſtles, and is thirteen lines long. Dark gray and white, intermixed with ſome ruſty lines on the wings, compoſe its varied plumage; white predominates on the under ſurface of the body, where it is ſprinkled with long blackiſh ſpots; the blackiſh, on the other hand, is the prevailing colour on the back, where the white forms only ſome edgings. Two white lines run obliquely, the one over the eyes, the other below them; ſmall blackiſh feathers half conceal the yellow ſpot on the crown of the head. The feathers of the tail, which are black in the middle, have broad borders of rufous; the hind nail is the ſtrongeſt of all.—The *Caudec* haunts the creeks, and perches on the low branches of trees, feeding probably upon aquatic inſects. It is leſs frequent than the *Titiri*, but has the ſame audacity and cruelty. In the female, the yellow ſpot is wanting on the head; and in ſome males that ſpot is orange, a difference which is perhaps owing to the age.

The TYRANT OF LOUISIANA.

Muſcicapa Ludoviciana, Gmel.

Seventh Species.

This bird was ſent from Louiſiana to the Royal cabinet, under the name of *Flycatcher*, but ought to be ranged with the Tyrants. It is as large as the Red-backed Shrike; its bill is long, flat, beſet with briſtles, and hooked; its plumage is gray-brown on the head and back, light ſlate-colour on the throat, yellowiſh on the belly, and light rufous on the great coverts; its wings cover only the third part of its tail, which is a brown aſh-colour, ſhaded with a little rufous from the wings. We are unacquainted with its inſtincts, but theſe features ſufficiently characterize it; and as it has the ſtrength of the Pipiris, it probably has alſo their habits. [A]

[A] Specific character of the *Muſcicapa Ludoviciana*:—"It is "brown-cinereous; below yellowiſh; its throat ſlate-colour; the "quills of its wings, and the edges of thoſe of its tail, rufous."

BIRDS

RELATED TO THE FLYCATCHERS, THE MOUCHEROLLES, AND THE TYRANTS.

The KINKI-MANOU of MADAGASCAR.

Muſcicapa Cana, Gmel.
Muſcicapa Madagaſcarenſis Cinerea Major, Briſſ.
The Aſh-coloured Flycatcher, Lath.

THIS bird is diſtinguiſhed from the Flycatchers by its ſize, being almoſt as large as a Shrike; but it reſembles them in many other characters; though a contiguous ſpecies, therefore, it cannot be included among them, but evinces that our artificial diviſions correſpond not to the diſcriminating lines traced by Nature. The Kinki-Manou is eight inches and a half long, and is bulky; its head is black; and that colour extends like a round hood on the top of its neck and under its bill; the upper-part of its body is cinereous, and the under-part aſh-blue; the bill is ſlightly hooked at the tip, and not ſo ſtrong as that of the Shrike, nor even ſo ſtrong as that of the Little Tyrant; a few ſhort briſtles riſe from the corner of the bill; the legs are of a lead colour, and thick and ſtrong.

The

The RED FLYCATCHER.

I am of opinion that the *Red Flycatcher* of Catesby, and the *Red Carolina Flycatcher* of Brisson, cannot be referred to the genus of the Flycatchers, or that of the Moucherolles; for though its size, the length of its tail, and even its mode of life, seem to be analogous, its bill is thick, large, and yellowish, which rather points its relation to the Yellow Bunting. We shall therefore regard it as an anomalous species. It is thus described by Catesby: "It is about the bulk of "a Sparrow; it has large black eyes; its bill "is thick, strong, and yellowish: the whole of "the bird is of a fine red, except the inner "fringes of the wing-quills, which are brown; "but those fringes are not seen unless the wings "are spread: it is a bird of passage, and leaves "Carolina and Virginia in the winter; the fe-"male is brown, with a yellow shade." Edwards also describes it, and admits, that it has the bill of the granivorous class, only *longer*. I think, adds he, that Catesby found that these birds feed upon flies, since he gives the Latin appellation of *Muscicapa Rubra*.

The

The DRONGO.

Lanius Forficatus, Gmel.
Muſcicapa Madagaſcarenſis Nigra Major Criſtata, Briſſ.
The Fork-tailed Shrike, Lath.

Though nomenclators have claſſed this bird with the Flycatchers, it appears to differ widely both from theſe, and from the Moucherolles; we have therefore ſeparated it entirely, and aſſigned it the name of *Drongo*, which it receives in Madagaſcar. Its characters are: 1. Its bulk, being larger than the Blackbird, and thicker: 2. The tuft on the origin of the bill: 3. Its bill is not ſo flat: 4. The tarſus and toes are very ſtrong. All its plumage is black, varying with green; directly under the root of the upper-mandible ſome long and very narrow feathers riſe erect to the height of an inch and eight lines; they bend forward, and make a very odd ſort of creſt; the two outer-quills of the tail project an inch and ſeven lines beyond the two middle ones; the others are of an intermediate length, and diverge, which occaſions the tail to be very forked. Commerſon aſſures us, that the Drongo has a pleaſant warble, which he compares to the ſong of the Nightingale; and this makes a wide difference from the Tyrants, which have all ſhrill cries, and are beſides natives of America. This Drongo was firſt brought from

 Madagaſcar

Madagaſcar by Poivre; it has alſo come from the Cape of Good Hope, and from China. We have remarked that the creſt is wanting in ſome ſpecimens, and we have no doubt that the bird ſent to the Royal cabinet under the name of *the Forked-tail Flycatcher of China*, belongs to this ſpecies, and is perhaps a female; the reſemblance, if we except the creſt, being entire between this Chineſe bird and the Drongo.

There is alſo a kind of Drongo found on the Malabar coaſt, whence it was ſent by Sonnerat: it is rather larger than that of Madagaſcar, or that of China; its plumage is wholly black, but its bill is ſtronger and thicker; it has not the creſt; and what the moſt diſtinguiſhes it are, the two long ſhafts which project from the ends of the two outer-quills of the tail; they are almoſt bare for ſix inches of their length, and have webs near their extremities as at their origin. We are unacquainted with the habits of this bird of Malabar; but they are probably the ſame as thoſe of the Drongo of Madagaſcar, ſince the external characters are alike in both. [A]

[A] Specific character of the *Lanius Forficatus*:—“Its tail “is forked; it has an erect creſt on its front; its body is greeniſh-“black.”

The

The PIAUHAU, *Buff.*

Muſcicapa Rubricollis, Gmel.
Muſcicapa Cayanenſis Nigra Major, Briſſ.
The Purple-throated Flycatcher, Lath.

The Piauhau is larger than any of the Tyrants, and is therefore excluded from the Flycatchers; indeed, except in its bill, it bears not the leaſt analogy to theſe, and ſeems to occupy a detached place in the order of Nature.

It is eleven inches long, and is larger than the Miſſel Thruſh. All its plumage is deep black, except a deep purple ſpot that covers the throat in the male, but is wanting in the female; the wings, when cloſed, extend as far as the end of the tail; the bill is ſixteen lines long, and eight broad at the baſe, very flat, and ſhaped almoſt like an iſoſceles triangle, with a ſmall hook at the point.

Theſe birds move in flocks, and commonly precede the Toucans, and always uttering the ſhrill cry *pihauhau:* it is ſaid that they feed upon fruits like the Toucans; but probably they alſo eat the winged inſects, for the catching of which Nature ſeems to have faſhioned their bill. They are very lively, and almoſt in continual motion. They reſide only in the woods like the Toucans, and generally haunt the ſame ſpots.

Briſſon aſks if the Jacapu of Marcgrave be not the ſame with the Piauhau *? We may anſwer that it is not. The Jacapu of Marcgrave has indeed a black plumage, with only a purple, or rather a red ſpot under its throat; but at the ſame time, *its tail is long*, *its wing is ſhort*, *and its ſize is that of a Lark*; theſe characters do not apply to the Piauhau. [A]

Thus the Kinki-Manou and the Drongo of Madagaſcar, the Red Flycatcher of Virginia and the Piahau of Cayenne, are all contiguous ſpecies, but eſſentially different from thoſe of the Flycatchers, the *Moucherolles*, and the Tyrants.

* "Jacupu, a bird of the bulk of a Lark; its tail extended; its legs ſhort and black; its nails ſharp on the four toes; its bill ſomewhat curved and black, half an inch long; its whole body is clothed with black ſhining feathers; but under the throat, ſpots of vermilion are mixed with this black."

[A] Specific character of the *Muſcicapa Rubricollis*:—" It is " black, a great ſpace on its chin and throat red."

END OF THE FOURTH VOLUME.

Printed in the United States
By Bookmasters